Mary Ellen's
1,000 NEW
Helpful
Hints

Books by Mary Ellen:

MARY ELLEN'S BEST OF HELPFUL HINTS
MARY ELLEN'S BEST OF HELPFUL KITCHEN HINTS
MARY ELLEN'S BEST OF HELPFUL HINTS BOOK II
MARY ELLEN'S HELP YOURSELF DIET PLAN

Mary Ellen's
1,000 NEW
Helpful
Hints
by
MARY ELLEN PINKHAM

Illustrations by Lynn Johnston

DOUBLEDAY & COMPANY, INC.
GARDEN CITY, NEW YORK
1983

The hints in this book are
intended to suggest possible
solutions only. The author and
publisher cannot guarantee
absolute success. To guard
against damage, we recommend care.

Library of Congress Cataloging in Publication Data
Pinkham, Mary Ellen.
Mary Ellen's 1,000 new helpful hints.
Includes index.
1. Home economics.   I. Title.
II. Title: Mary Ellen's one thousand new helpful hints.
III. Title: 1,000 new helpful hints.
IV. Title: One thousand new helpful hints.
TX158.P558   1983   640

ISBN: 0-385-18963-x
Library of Congress Catalog Card Number: 83-45112

1,000 New Helpful Hints For:

Introduction

People tell me that when I demonstrate hints on TV I seem to be having lots of fun. Maybe that's why every once in a while I see a few familiar TV faces trying to get into the act. One night I saw David Letterman pass along several "hints" to his late-night audience. He told them to take those drops of toothpaste that harden on your sink and serve them as after-dinner mints. He suggested that they not fry hamburgers in the toaster. And he hinted that you could make emergency draperies out of a Scarlett O'Hara-type ball gown.

Well, he was kidding, but when I tell you to put your angora sweater in the freezer or spread shaving cream on your carpet, that's no joke. Freezing the sweater keeps it from shedding, and shaving cream actually cleans the rug.

Some of my hints sound funny, but, in fact, they work!!!

Since I started "hinting," I've collected the best solutions to all kinds of household problems that come up when you do things like clean, paint, garden or care for children. You may have read and (I hope) enjoyed them in my three other books. Now here's a fourth book with some of the best tips yet.

I know that some of you will turn right to the sewing section and others will be looking for ideas about entertaining. My readers have many interests and concerns, and I've tried to share ideas for as many of them as possible. I figure that if you find just ten hints in this book that make your life easier or more interesting — and maybe save you some money besides — I'll feel I've done my job. Please let me know.

Helpfully,

Mary Ellen

1

AUTO MAINTENANCE

You Auto Know About These

Cleaning vinyl seats
- When vinyl seats in your car get grimy, spray on foaming bathroom cleaner and work into the seats with a bristle brush. Wipe with a damp sponge. This works especially well on textured vinyl.
- Or, use a powdered hand soap containing borax, such as Boraxo. Just dampen a sponge and apply soap in a circular motion to seats. Rinse well.
- Or, wipe vinyl seats (and dashboard) with a cloth dampened with self-stripping floor wax cleaner. It removes dirt, covers scratches and scuffs, and dries quickly to a nonstick shine.

Cleaning the dash
- Use a sponge-tip paintbrush that's been dampened to clean hard-to-reach areas.

Car body cleaner
- A grease-cutting liquid dishwashing detergent will help remove built-up grease and grime from the body of your car. It really works!

Cleaning white walls
- If they're scuffed up, use soap-filled scouring pads.
- Or try one of the prewash sprays and a little elbow grease.
- Once the tires are clean, rub on a little vinyl top wax. They'll look clean longer.

Great brush idea
- Use a baby bottle brush to clean those hard-to-reach places on tire-wheel covers.

Cleaning a steering wheel
- Keep a box of moist towelettes in the glove compartment for fast cleanups.

Did you know?

● The heavier the car, the more fuel it takes to run it. For every one hundred pounds of additional weight in the trunk, gas mileage will be reduced by 2 percent.

Gas caps

● You won't forget your gas cap at self-service gas stations if you attach a magnet on the inside of the door to the tank. The magnet holds the gas cap while you fill the tank. Since the door won't close until the cap is properly in place, you can't forget it.
● Or, attach the magnet to your car's license plate.

Oil's well

● When removing an oil filter from the car, place a strip of sandpaper between the wrench and the oil filter to keep it from slipping.
● Keep open cans of motor oil clean by sealing them with the plastic lids from one-pound tins of coffee.

Do-it-yourself oil change

● Partially loosen the filter, allowing the oil to drain from it. Then, place a small plastic sandwich bag over the filter and finish unscrewing it. Now, all the mess is in the bag.

Removing car grease from hands

● Wash with Ivory dishwashing liquid and water. It cleans like a charm.

Choosing the right gas

● If your car "knocks," don't go in for a tune-up until you try a higher-octane gasoline. More than likely, that'll calm 'er down.

Unfreezing a door lock

● Cup your hands around the lock and breathe on it. The lock will open every time.

Defogging the inside windshield

● Use an old windshield wiper blade on the inside of the window.

Extra storage from your garage

● Lay a platform across the ceiling joists and use the space between the ceiling and the roof for storage.
● Or, shelves or cabinets can fill the top half of your garage's front wall, since the hood of your car doesn't occupy that space.
● Or, old chests of drawers come in handy in the garage if you've extra floor room. Label each drawer so you know what's stashed inside: battery cables, extension cords, garden tools or whatever.

Trouble on the Road

If your gas tank springs a leak
● Plug it up temporarily with a wad of chewing gum.

Light the way
● Find it hard getting light just right to see what you're repairing? Attach a bicycle mirror to a flashlight so the beam of light will be directed at an angle, lighting up the area. Seeing "around corners" will be a big help.

Let it snow
● Expecting snow? Park your car at the end of the driveway near the street. You won't have to shovel yourself out before leaving.
● Keep a large bag of rock salt in the trunk to add weight to the car and improve traction in snowy weather.

Two more "bright" ideas
● Put a strip of bright reflector tape on some old coffee cans and store them in your trunk. They make great safety lights in an emergency.
● Keep a bright-colored ribbon or bandanna in the glove compartment. Pull out and tie to the antenna when you're stalled in a blizzard and want to be noticed — or even to avoid "losing" your car in the parking lot during a shopping trip.

BEAUTY AND HEALTH

Hints for All of You

File this one away
- Never file nails after washing dishes, showering or bathing. Nails that have been softened from hot water will split or peel. Let them dry before doing any touch-ups to smooth them out.

You'll wonder where the yellow went
- Remove the yellowing stains you get from nail polish by brushing toothpaste onto nails, then rinsing them off.

Another tip for toothpaste
- Ballpoint ink or adhesive residue marking up your nails? Rub toothpaste on and watch the stain vanish.

Trying out nail polish
- Here's how to do it without disturbing your present manicure. Put on a light coating of hand cream, then wrap a piece of cellophane tape on your fingertip. Paint on the new nail color. It's a great way to audition a color and see how it will actually look on you.

Longer lasting polish
- Dip your fingernails into a solution of one-half cup of water and two teaspoons of white vinegar. Pat dry, then apply polish. You'll get an extra bonus, too. Vinegar softens cuticles and conditions nails.

In the chips
- Freshen chipped color without redoing your nails. Lightly dip a cotton swab into nail polish remover and rub the swab over the chip to soften the edges. Then repolish the tips.

Quick nail drying

- If you're out of nail drying product, spray nails with a cooking oil like Pam. In five minutes, lightly tissue off the excess.
- Or, let polish set a few minutes, then run cold water over nails.
- Or, dip freshly polished nails into a bowl of water and ice cubes for a quick dry.

When nail polish gets clumpy

- Add a few drops of rubbing alcohol to thin it out.
- Or, a few drops of nail polish remover.
- Or, cap bottle tightly and run hot water over it. It will then be easier to apply.

Guidelines for cutting bangs when hair is wet

- If hair is straight, cut bangs at the bridge of the nose.
- If hair is wavy, cut one-quarter inch below the bridge.
- If curly, cut halfway down the nose.

Hair conditioner to make yourself

- Add one teaspoon of gin or vodka to two beaten egg yolks. Wash your hair as usual, and towel dry. Then massage the mixture into scalp and hair and leave on for five minutes. Rinse thoroughly.

Great setting lotions
- Club soda straight from the bottle.
- Or, flat Budweiser.
- Or, mix one teaspoon of unflavored gelatin to two cups of boiling water. Cool, then pour into a spray bottle and use for a great set.

For a faster set
- Put your brush or plastic rollers in a mesh vegetable bag, dip in a basin of warm water and shake the bag vigorously over the tub. Set your hair with the slightly damp rollers and it'll dry in a short time. A good way to spare your hair excessive hot roller treatment.

Permanent solutions
- Make sure hair is completely saturated with waving lotion by pouring the lotion into a clean pump bottle and spraying it on. It's a lot more thorough than dabbing away at it by hand.

Saving your skin
- Put petroleum jelly around your hairline to keep hair dye or a home permanent solution off your face.
- Place a cotton strip over the petroleum jelly on your forehead and you'll avoid drips, too.

Saving the sink from dye drops
- When using henna or color on your hair, coat the sink lightly with liquid dishwashing detergent. No sink stains when dye drips or runs.

Fighting fire with fire
- If your tint has stained your skin around the hairline — or anywhere else — guess what! Hair color removes hair color stains. Just rub leftover tint into the stained area and rinse with water.

Protecting your hairdo
- Wrap a large scarf around your head and gather the four corners in your mouth. Slip the sweater or dress over your head.
- Or, pull a half slip over your head and slip clothes over it.

Sunblock for scalp
- Scalps burn too. Apply sunblock to the part of your hair with a Q-tip if you expect to be out in the sun without a hat.

Keeping track of coated elastic bands
- When you buy them by the package, wrap all of them around the handle of your hairbrush. You won't have to go searching for the package or the loose bands.

Keeping track of pics and clips
- Push roller pics through a plastic pot scrubber.
- Hang electric roller clips over the rim of a mug. No more tangles, and the clips hang neatly.

Washing foam rollers
- Put foam rollers in a lingerie bag and wash them on the gentle cycle with the rest of the laundry.

Washing hair curlers, combs and brushes
- Put them in a nylon bag and toss into the washing machine.
- Or, add two or three tablespoons of household ammonia to a basin of hot water. Scrub with a nailbrush if curlers, combs or brushes are especially grimy.
- Or, clean combs with a prewash like Biz or Axion. Sprinkle the prewash (dry) into a basin of water. When it's dissolved, toss in combs and let them soak until clean, then rinse.

How dry I am
- To eliminate the annoyance of water running down your arms or soaking long sleeves while you wash your face, wear wrist sweatbands.

- Or, use a head sweatband wrapped twice around your wrists.

Getting the yellow out of a tan
- If your tan is fading and skin has a yellowish cast, use a moisturizer or moisturizing makeup with a pinker tone.

Covering bathing suit strapmarks
- If you've changed swimsuit styles and your upper torso looks like a patchwork quilt, use sunblock on tanned areas (to keep them from getting darker) and use a quick-tanning lotion on pale ones. With a little luck, you should be an even color in a few days.

Reducing large pores
- Put a pinch of powdered alum (available at the spice section at the grocer's or the drugstore) into skin fresheners, toners or astringents, then shake and apply to skin. Large pores should be less noticeable.

Milk of magnesia masque
- Getting an upset stomach because of your oily skin and blackheads? Take out the milk of magnesia—but apply it to your *face*. Leave on a few minutes, then rinse off. Use only once a week. Trust me: This really works.

Pointers for makeup pencils
- If your brow, lip or eye pencils or crayons are too soft, they'll break when sharpened. Chill first in the fridge.
- If your makeup pencil is too scratchy, warm the point between your fingertips to soften before using.

A matter of mascara
- Don't "pump" your mascara wand with each application. That forces in air, which dries up mascara. Instead, just pull the wand straight out and apply.
- If mascara does dry out, thin with a drop or two of liquid glycerine from the drugstore. Or try running hot water on the outside of the tightly capped container.

- If mascara is too dry to use on your lashes, don't throw it out. Brush lightly on eyebrows instead of using eyebrow pencil.

Smoother looks with eye shadow
- Use cream and powdered eye shadows in the same shade when trying this: Put the cream on first, then dust with the powdered shadow. Now it will stay put.

Shining braces
- If you wear braces, mix equal parts of salt and baking soda and brush with this mixture. Bands and wires will be nice and shiny. Be sure to smile and wear them proudly.

A better pedicure
- Since pedicures are best if feet are soaked first, put the stopper in the drain when you take a shower. Feet can soak while you shower.

Prevent snagging of stockings
- Rub lotion into feet and legs before you smooth on stockings.

Shaving legs
- Prevent friction burns when shaving legs with an electric razor by dusting legs with baby powder first. Clean razor after shaving.

To remove tar from the bottom of your feet
- Nothing works better than toothpaste. Tar comes right off.

Eliminate makeup stains on clothes
- So you don't smear makeup and stain clothes at the same time, stretch a hairnet for bouffant hairdos over your face. Now put on clothes. Keep this net handy for makeup only.

Tools of the Trade

Clogged spray cans
- Remove the tip from your deodorant or hair spray and hold under hot running water.
- Or, borrow a new tip from another can.

Cosmetic sponges
- Keep them bacteria-free. Rinse and store them in plastic bags in the refrigerator.

Substitute for a pumice stone
- Try the abrasive side of a Scotchbrite Sponge. Some feel it does an even better job.

Cleanups with old toothbrushes
- Save worn ones to clean blades on double-blade razors. If you brush out the stubble that's caught between the blades, you might be able to get another shave or two out of them.

A tight squeeze
- If tweezers are dull, lightly go over the tips with a nail file, emery board or pumice stone. Tweezers will be as good as new.
- Just use an old-fashioned clothespin, slip it on and use as a key to roll up a tube of toothpaste. A great way to squeeze the last bit out.

All Steamed Up

In the sauna or steam room
- Get the most out of it: Moisturizers won't allow sweat to flow freely from your pores, so be sure not to apply any to face or body.

- Use a moisturizer after showering.
- Try giving your hair a conditioning treatment. Heat from the sauna or steam room will help conditioners to penetrate hair shaft.

After your workout
- You don't need extra salt after exercising, but you need to replenish liquids lost from perspiring. Pass up the salt tablets, potato chips or pretzels (which add calories, too), and sip water instead.

Taking Care

First aid for the medicine cabinet
- Glue small magnets inside the medicine cabinet for holding files, cuticle scissors and other metal objects.
- Organize medicine bottles in old ice-cube trays, lipsticks in spray-can lids and just about anything into small plastic containers.
- Strip a piece of adhesive tape onto the front edge of the shelves to keep medicine bottles from slipping off.

Your own emergency kit
- Use a plastic file card box and fill with essentials for first aid. Tape a list of the contents and their expiration dates inside the lid to remind you if you need replacements.

Substitute hot water bottle
- Use a large, empty plastic bottle from dishwashing liquid. Fill with hot water, screw the cap on tightly, wrap a dish towel around it and you've got your bottle.

Cure for a leaky ice bag
- Fit a plastic bag into the ice bag, fill, and tie the plastic bag closed before you cap the bag. You've got an ice bag within an ice bag and no leaks.

Butterscotch swirl gone to your head?

● If you get a headache shortly after eating ice cream, it's because the cold chills your palate and radiates up to the brain. Eat ice cream slowly. Give your mouth time to adjust to the change in temperature and you'll be less likely to get a headache.

Water near the brain?

● Here's a great way to get rid of plugged ears from swimming. Tilt the ear with the water in it toward the floor, then jump up on the corresponding foot — right ear, right foot; left ear, left foot. The ear will immediately unplug. Don't laugh. It works!

Tilt!

● If you can't read the numbers on your scale, just put a magnifying glass over the dial. Even if you don't believe the numbers, at least you'll be able to see them.

Cutting calories
● Tomato juice is not only lower in calories than apple or orange juice, but the acidity may even help curb your appetite if you drink it before your meal.

Zzzz!
● People snore only when they're lying on their backs — so keep your snorer on his side or stomach this way. Put a marble in a piece of cheesecloth and sew it to the back of his pajamas between the shoulder blades. When he rolls on his back, the marble will wake him and he'll change position.

Don't Take a Bath on Bathroom Accessories

Uses for plastic bottles
● Fill with shampoo for kids.
● Use for hand lotion. Pour into the bottles from large economy sizes.
● Fill with hair-setting lotion

Saving on soap
● Keep a pint jar with a half cup of water in the bathroom. Drop all scraps of bath and hand soap into it. When the jar's full, run the contents through a food processor for thirty seconds. Pour the liquid soap into a pump-type dispenser and you've got a constant supply of soap. Just rinse the processor in hot water to get all the soap out.

Soap lost its suds?
● Just prick it deeply with a fork and you should get as much lather as when the bar was new.

No more waterlogged tissues
- If tissues repeatedly get soggy on a damp shelf or counter, you don't need to buy a dispenser. Instead, put little "legs" on your tissue box. Pushpins (the kind that hold messages on bulletin boards) do the trick and come in colors that will match any box.

Seeing Is Believing

Making contact
- To find a dropped hard contact lens, try this: Darken the room and shine a flashlight over the floor. The lens will sparkle in the light.

Cleaning lenses
- Drop them in a sterilized spice jar that has a perforated plastic lid. You can wash the lenses, drain and rinse them in a few motions.

For your loose screws
- If the screws on the bows of your glasses tend to work loose, touch the top of them with a drop of clear nail polish.

Scent-sations

Getting your scents' worth
- For best results, spray scent on the inside of your wrist at the pulse point and rub it in. Wait a minute, then sniff. If you like the scent, shop for a while, then sniff the scent again to see if the chemistry's really right before you buy it.

Fragrance for less
- You can save money by buying your favorite scent in a bath oil rather than the cologne or perfume. Dab it on just like perfume. Bath oil is absorbed into skin and the intensity of the fragrance is nearly the same.

Save more on fragrances
- Buy large bottles of cologne and transfer some to a clean pump-spray bottle. Spray-on perfume is easier to apply and much more fun to use.

Longer-lasting fragrances
- Apply when skin is still warm and slightly damp after bathing.
- Or, soak a cotton ball with the fragrance and tuck it into your bra.

Scenting clothing
- Instead of spraying it directly on clothing, lightly spray the ironing board, then press clothes.

Scents in the sun
- Don't dab perfume directly onto your skin before going into sunlight, since some ingredients may cause pigment spots to appear. Instead, apply it to hair, inside your wrist, or on clothes it won't stain.

For people who are sensitive to perfume but want to wear it
- Spray perfume around the inside heel of your shoes. The heat from your feet will help release the lovely scent. You shouldn't have any problems about direct contact with the scent since you're wearing panty hose.

Perfume care
- Be sure bottles of perfume and cologne are out of direct sunlight. Sun can cause evaporation and, in some cases, change the fragrance.

Good to the last drop

- Keep the bottle when perfume runs out and set it un-opened in a lingerie or sweater drawer. Your clothes will smell sweet — just like you.

Reusing perfume bottle

- If you want to keep the bottle the cologne or perfume came in, fill it with rubbing alcohol and let it stand overnight. Rinse out, then wash with soap and water.
- Or, use baking soda and warm water. Let stand overnight.

CHILDREN

Open Up Wide

Introducing solid food
- It's best to introduce one new food a week, never more than that. In case the baby has any allergic reactions, you'll always know exactly which food is causing it.
- Rice cereal will be easier to give baby if you mix it with the formula and run it through the blender. When it's all liquefied, pour into a bottle. The cereal should flow right through the nipple. If it doesn't, enlarge the hole.
- The baby will be more likely to eat a new food when he's the hungriest so feed it to him at the start of a meal.

Grinding food
- Soft-cooked meat will be easier to grind in the baby-food grinder if you add a little liquid to it.

Spoon-fed
- While you feed baby with a spoon, give him a spoon of his own to play with. Before long, he'll learn to feed himself by copying you.
- Mashed potatoes are a good training food for handling utensils since they don't fall off a spoon easily.

"Bib-bits"
- If you tuck a sponge in the pocket of a plastic bib, it'll stay open wide enough to catch any spilled liquids.
- Safety-pin the bottom of the bib to baby's clothes for lots better coverage.

Sitting pretty in the highchair
- Use bath towels to pad the back of the highchair. The towels will also add back support while he sits.
- Or put a thick foam rubber cushion on the highchair seat. The cushion also helps keep spilled liquids from dripping onto the floor.

- Or place a small rubber sink mat on the seat.
- Put newspaper down underneath, instead of plastic. You'll save time by throwing away the mess instead of scrubbing it up.

Off the bottle
- If baby refuses to drink from a cup, try this: Take the cap and nipple off the bottle and see if he'll drink that way until you can reintroduce the cup.
- Or give him his favorite drinks in the cup and his least favorite drinks in the bottle. It won't be long before he'll want everything in the cup.
- Remove that chalky residue by boiling baby's bottles for ten minutes in a pan of water to which one cup of white vinegar has been added.

Growing Strong and Healthy

Getting kids to eat vegetables
- If your kids don't reach for raw carrots and celery, maybe it's because they're not within immediate view. Try keeping cut-up celery and carrot sticks in the refrigerator in a glass of water. Vegetables will stay crisp and available.
- Or take the labels off several canned vegetables before storing them in the cupboard and let kids pick which of the mysterious, unmarked cans to open.

Healthy drink dispenser
- Keep a picnic-size jug with a spout filled with your children's favorite pure juice drink and place it on the kitchen counter. Leave a supply of paper cups nearby so the kids will be tempted to reach for the juice—not for the soda pop.

Making It All Better

Teething treats
● For a soothing treat while baby's teething, let him suck on a frozen banana.

Swallowing medicine
● Give your child an ice cube to suck on before giving her bitter-tasting medicine. Her taste buds will be desensitized and medicine will go down a bit easier.

When it's a bad dream
● A toddler may wake up crying from what seems to be a nightmare. To comfort him, just wipe his face with lukewarm water and a wash cloth until he knows what's going on. He'll stop crying and be able to fall right back to sleep.

Removing gum from hair
● Apply Spray 'n Wash generously to the gum, then rub hair between fingers. Comb the gum out, then wash the child's hair. It beats whacking it off with the scissors!

Saving Band-Aids
● Kids go through Band-Aids at a rate that can make your head spin. Cut down on waste by keeping only a few of them in the box and the rest out of sight.

Out of sight, out of mind
● If your oven controls are located at the front of your stove within toddler's reach, pull off the knobs and leave them off when not in use. When using the stove, set the temperature, then remove the knobs again so toddler can't play with them.

Wading pool safety
● Unless you're there to keep an eye out, empty the pool so little kids won't be tempted to jump in.

Be prepared!
● Play the "what if" game with kids. Give them situations to solve, like getting separated from you at the department store and ask them what they would do. If they answer correctly, praise them. If they don't know what to do, here's the perfect chance to talk things out with them — for their safety and your peace of mind.

Doing It the Right Way

Telling right from left
● Do your children have a hard time remembering right from left? Just have them extend the thumb and forefinger of both hands. The hand that makes an "L" is the left one.

AMY SAUNDERS
126 MAPLE AVE.
I'M 5 YEARS OLD.....

Remembering address and phone number
● Try setting both to music. Kids will remember a lot easier if important information is set to a tune they can sing.

Practicing writing names
● A preschooler can learn to write his name correctly this way: Write each letter of the name in dot-to-dot fashion on a small piece of cardboard. Cover with clear contact paper so your child has a washable surface to practice on.

Putting on socks
● Your toddler can learn to put on his own socks if you start him off with tube socks. Since there's no shaped heel, your child can't put them on wrong.

Hanging up jackets and hats
● Nail an accordion-type mug rack at child height by the entryway. See the difference it makes!

Sitting still
● Is your toddler especially squirmy when you try to dress him? Sit him in the highchair with a small toy to hold. It'll be lots easier to put socks, shoes and shirts on him.

Avoiding shampoo tears
● Rub hair with a lathered washcloth, then rinse with a washed-out cloth.

Brushing teeth
● If you don't have time to watch the kids brushing, invest in a three-minute egg timer and tell them they have to keep brushing until the sands run out. Children think the timer's fun and you know they're not rushing the job.

Filling the tub
● If big kids want to fill the tub to the top, prevent quarrels by sticking a decal on the side of the tub. When water reaches the decal, they'll know it's time to turn the water off.
● Be sure you turn off the cold water last before putting a small child in the tub. In case the faucet drips, fingers and toes won't be burned from hot water.

Organizing bath toys
● Purchase a plastic bicycle basket. The handles fit perfectly over the bathtub soap dish. Toys can drain overnight since the basket has an open weave.

Diapering is a snap
● New mothers and baby gift-givers should keep in mind that pants and overalls with snap bottoms are more convenient when changing diapers.

Weighing in with baby
● Here's how to watch your own weight while keeping track of baby's. Get on the bathroom scale alone, then weigh yourself while holding baby in your arms. Baby's weight is the difference between the two.

Hanging posters
● Try using straight pins, tapping them halfway into the wall. Posters are easily moved, since the pins leave holes that are nearly invisible.

Storing books
● Use plastic napkin holders for small hardcover storybooks.

Using every bit of closet space
● Put in shelves across the bottom half of the closet and use for foldable clothes like pajamas or underwear, or stack games and toys.

Finding clothes faster
● If two kids of the same sex share the same closet, mark hanger tops with strips of tape in different colors.
● Or buy each child hangers in different colors.

Funstuff

Pushbutton birthday song
- Play the "Happy Birthday" song on a Touch-tone phone while on the line with a friend by pushing:

  1 1 4 1 * 9
  1 1 4 1 # 0
  1 1 # 0 * 7 4
  # # 0 * 0 *

Thank-you note from baby
- Take a snapshot of baby wearing or playing with a gift from a relative and send along with a thank-you note.

Bonus gift
- A great wrapping idea for a birthday present: a jump rope tied around a box and made into a bow.

Picture this
- Tape a long sheet of brown mailing paper or shelf paper to the floor. Hand out crayons and let each child draw a picture on his own section of the sheet. When the party's over, cut out each child's section so he can take it home.

The best toy box
- An open-weave plastic laundry basket is great. The kids can see what's in the basket, besides having lots of room to store stuff.

Portable toy storage
- Attach a cord handle to a two-gallon ice-cream bucket and hang it over one handle of your toddler's stroller.

Paint containers

- A styrofoam egg carton makes a handy container for young finger painters. Use a different well for each color and when they're through, close the lid and throw it away.

Roll-on "paintbrush"

- Pry the top off a roll-on deodorant bottle and rinse both parts. Fill the jar with thinned finger paint or poster paint and replace the top.

Mail call

- You feel left out when your mailbox is empty. So do your kids. Buy them subscriptions to children's magazines and they'll love you for it.

Ride 'em, cowboy

- Convert your broom into a "horse" by stuffing an old sock and tying it to the broom handle. Draw on eyes and a mouth and the kids will be ready to ride on it.

Let's get organized

- Glue a manila envelope to the inside top of each of your game boxes. It'll keep rules, order forms and score pads and pencils that come with the game all organized.

Table tennis caddy

- Use an ordinary clipboard to keep paddles and balls in one place. Drop balls into a plastic bag, then slip paddles and the bag of balls under the clip.

And Away We Go

At the show

- It's always a good idea to bring a kiddie seat with you when you take the kids to the movies, ball games or other sit-down events. Kids won't miss any of the action this way.

- If you're leaving a movie theater or sports stadium or any place where there's a huge crowd trying to get out at once, veer left. Most people automatically bear right. You'll be fighting through fewer people in the opposite direction.

At the restaurant
- Ask the waitress to bring your toddler's dinner about three or four minutes before yours. That should give you time to cut meat, mash vegetables and settle toddlers in for dinner. When your food is served, you should be able to eat in peace.
- Don't get caught without a bib for very young children. Just carry a sweater chain in your purse and use it to clip a napkin around your child's neck.

At a friend's house
- If you're visiting with a very young child, protect your friend's bedding when putting him down to nap. Double a sheet of aluminum foil and put it under the sheet or towel where baby will be napping. The foil keeps bed dry while it warms to body temperature.

On the bus
- If kids take public transportation, make packets of the exact change and keep them in a basket by the door. They can just grab a packet on their way out and never be caught without the exact change.

CLEANING

Quick Cleaning Guide

Black lacquer
- Wipe on a strong solution of tea with a soft cloth and wipe dry to a nice luster.

Books
- Remove nongreasy dried spots from book pages by rubbing them *very* gently with extra-fine sandpaper.
- When a book gets wet, try this: Dry the pages and keep them from wrinkling by placing paper towels on both sides of every wet page. Close the book and put a few heavy books on top of it overnight.

Brass
- Apply a thin coat of window cleaner with a soft cloth, let dry and rub lightly to polish. Brass will be tarnish-free for months.

Bricks
- Squeeze some denture cream onto a wet toothbrush and scrub the grout. Wipe clean with a damp cloth.

Brooms
- Try a wide-tooth comb for cleaning the fuzz from broom bristles.

Charcoal grill
- File a notch in the blunt end of a beer can opener and use it to scrape the grill.

China
- To get nicotine stains off favorite china dishes, rub the spot with a cork or soft rag dipped in baking soda

Chrome
- Nail polish remover is great for cleaning chrome decorations and knobs, especially on stoves. Be sure all electrical or gas units are turned off, and rinse well with water.

Cobwebs
- Clean with an upward motion to lift them off. Downward motions tend to splatter them against walls.

Curtains
- Freshen them in the dryer with a fabric softener sheet and a damp towel.

Dish drainers
- Remove hard water stains by tilting up the low end of the board slightly and pouring one cup of white vinegar over the board. Let it set overnight and rub off with a sponge in the morning.

Door
- Is your child sticker-crazy? Hang a large sheet of clear plastic adhesive-backed paper on the door or wall before he puts any up. Stickers just peel right off when he tires of them.

Dusting
- Use a child's dust mop for dusting hard-to-reach places such as above doors, paneled walls, etc.
- Or, attach an old sock over a yardstick and secure with a rubber band. It comes in handy for cleaning cobwebs off the ceiling, too.
- A four-inch paintbrush is good to dust lampshade pleats, windowsills, etc.
- Spray some furniture polish on the bristles of your broom and the dust and dirt will be easier to collect when you sweep.

- Wipe miniblinds or vertical blinds with damp fabric softener sheets to eliminate static that collects dust. The same trick works for your TV screen.

Fire screen
- Vacuum off the dust, then scrub the screen with a sponge and suds. When the screen is dry, brush on a coat of self-polishing floor wax.

Fluorescent bulbs
- Here's a simple, safe way to dispose of them. Wrap them in overlapping newspapers, then secure with masking tape. Tap lightly to break the bulb, then discard in garbage.

Fringed rugs
- Keep fringes from getting caught in the vacuum by tucking them under the rug.

Garbage
- Drop the twist tie in the bottom of any can or basket before you put the liner in. When you're ready to empty it, the tie is handy.
- To prevent paper garbage bags from ripping down the side, turn the top edge of the bag down two inches to form a double edge. Bags won't collapse when you toss in the garbage.
- Drive a stake or broom handle into the ground and slip the side handles of the trash can over it. Cans won't tip.
- Use a child's skateboard to roll the trash can to the curb on pick-up days.

Gutters
- To remove leaves, start at the opposite end of the downspout. Then flood the gutter with water to help loosen the debris. Scrape, then flood again with water.

Heat registers
- Before you have the furnace "kick on," vacuum the heat registers so you're not hit by a cloud of dust.

Knives
- Got a trusty but rusty old knife that won't come clean? Sprinkle an abrasive cleanser on a cork, then dampen lightly and rub the cork against the knife. The cork's smooth surface makes for complete abrasion and the knife will look terrific.

Oven
- Before cleaning the bottom of the oven on an electric stove, prop up the heating element with a few clip clothespins. It will be held high enough so you can get underneath to clean.
- For easier jobs, keep a spray bottle filled with one-half ammonia and one-half water. Spray the oven walls, close the door, wait until the dirt softens, then wipe clean. Your hands will never touch the ammonia.

- After cleaning with a commercial cleaner, dissolve all traces of grease with a paste of baking soda and water. Spread on the oven floor and walls. Wait a second, then wipe clean.
- Eliminate the odor after oven cleaning: Place orange peelings on a rack in your oven and turn it up to 350° for a few minutes.

Rubber gloves
- Before using new gloves, turn them inside out and paint the fingertips with a few coats of nail polish. Your nails are less likely to poke holes.
- If they do, patch small holes and tears with waterproof tape. It works fine and cuts down on replacing gloves because of leaks due to holes.
- Out of rubber gloves? Make a substitute pair by slipping on plastic bags and securing them with rubber bands. Great for last-minute cleanups.

Shower doors
- A coat of acrylic floor finish gives new shine to fiberglass shower doors. Makes water spots disappear, too.

Sinks and tubs
- Hard water deposits: Use those fine pumice stones that are used to remove mineral deposits from the sides of swimming pools. Pick one up at a swimming pool supply store.
- Preventing rust stains on fiberglass: Wash and dry the tub well, then spray with a silicone spray and let dry before using. This treatment will protect the tub from rust deposits for at least a week.
- Nothing is better than an old toothbrush for crevices on a sink.

Silver
- Old powder puffs that have been washed and fluffed in the dryer are great for cleaning silver.

Storage
- Store the steel pad you're using in a little clay flowerpot. Clay absorbs moisture and helps prevent rusting.
- Use old wine racks to hold cleaning solutions. Just be sure to label the bottle caps and store in a safe place.

Tablecloths
- Drop soiled tablecloth and napkins in a sink filled with warm water and a half cup of Calgon water softener. Let stand until morning. In most cases, stains will vanish.

Toilet
- Drop a denture cleaning tablet into it for a fast cleanup!
- Or, use baking soda.
- If you'd rather clean it dry, get the water out this way: Pour a bucket of water in the bowl. This makes the water flush out, but the bowl won't fill up again. Clean away!
- Sometimes moisture accumulates around the toilet, leaving puddles on the floor. Prevent the condensation by applying a coat of floor wax to the tank.

Tub
- If your tub is extra dirty, mix a solution of automatic dishwater detergent and hot water. Let sit for twenty minutes, swish it around, then rinse with cold water.

FFWPPPP"

Vacuuming

- Put petroleum jelly on the ends of the vacuum cleaner extension wands and they'll slide apart easily.
- Prevent a retractable vacuum cleaner cord from snapping back into the machine when you don't want it to. Clip a spring-type clothespin to the cord pulled to the length you want it, and adjust as needed.
- No need to replace your fabric vacuum-cleaner bag when it tears. Simply mend it by pressing iron-on patches over the hole.

Vases

- Soak tea leaves in warm water. Pour the undrained mixture into the vase and either cover the top with your hand and shake it vigorously or just swish it around. Flush with cold water.

Venetian blind tapes
- Keep tapes from puckering or shrinking when washed by securing them top and bottom with adhesive-backed tape while they dry.

Vinyl floors
- For a shiny finish on no-wax floors, add about one quarter of a cup of a presoak laundry powder to soapy water and then mop.

Walls
- Cut out the toe from a discarded sock, roll the top over several times, and slip the sock onto your wrist. It'll soak up drips when you're washing walls — or woodwork or a ceiling.

Whisk broom
- Make one out of a sturdy vegetable brush.

Windows
- Remove stickers and decals by soaking with warm water, peel off what you can, then get the remaining sticky residue off with nail polish remover.

Wood
- Eliminate paint stains by rubbing with a cloth saturated in liquid self-polishing wax.
- Or, rub lightly with 3/0 steel wool dipped in a liquid polishing wax.

CLOTHING

Shopping with Savvy

Sales days
- The best shopping days are the first and last. There's often a better selection of merchandise on the first day and prices are usually lower on the last.

Finding less expensive casual clothes
- Look for them in departments that specialize in work clothes rather than in the higher-priced sportswear sections.

A shopping bag
- If you're going to be doing a lot of shopping, be sure to take a large trash bag along to keep packages together.

Buying an umbrella
- Hold it up to the light, opened. If you can see through it, you may have moisture seeping through it during a downpour.
- Count how many tacks hold the fabric to each rib. Two or three tacks mean a better grade than just one tack. Also: the greater the number of ribs, the sturdier the umbrella.
- Folding umbrellas without that automatic opening button feature are very flexible and more windproof.
- Umbrellas that open automatically are more likely to flip inside out than the standard ones.
- But, before you discard an old umbrella, remove the little metal tips that fit over each rib and use them to repair any umbrella that's lost one.

Buying bridal gowns
- Check the yellow pages for stores that sell discount, resale, sample or antique dresses. Don't overlook rental gowns either. Savings on the dress that's almost perfect can be 40 to 80 percent.

- Check the label and see where the gown is manufactured. If it's made locally, call the manufacturer's showroom and ask if they're open to the public. You can often save about 20 to 30 percent on direct showroom sales.

Stocking shades
- Test stocking shades this way before buying: Pull the hose over the inside of your forearm, not the back of your hand. Color will be truer since your forearms (like your legs) have less exposure to the sun. Since fluorescent lights in shops may add a greenish blue cast, if possible try to test colors in natural light.

From Your Head to Your Toes

Keep your hat on
- Have a hard time keeping hats that have a shallow crown on your head? Sew small grip-tooth combs to the sweatband.

Fitting hats snugly
- Keep one-size-fits-all hats on this way: Measure a piece of elastic around your head and cut to a snug fit. Weave it through the hat beneath the hatband so it can't be seen.

Freshening ribbons
- After washing and rinsing hair ribbons, dip them in a cup of warm water mixed with a teaspoon of sugar. Press out moisture on a towel, then iron while damp. The sugar water really gives ribbons body.
- To press a wrinkled hair ribbon, pull it across a hot bare light bulb. This works great with velvet ribbons that an iron would crush.

Drying ribbons and ties
● Wind them tightly around a wide glass jar instead of ironing them. They'll dry nice and smooth. Don't worry about keeping a tie in place on the jar—the ends will stick to each other.

Cleaning diamonds
● Nothing is better than Prell shampoo and a toothbrush for a quick cleanup.

Cleaning velvet
● The best velvet brush is another small piece of velvet rubbed *down* the nap of the garment.

Cleaning black lace
● Dip a sponge into cold tea and just dampen, don't soak. Put a piece of brown paper over it and press with a warm iron.

No more lost scarves
● Just sew a small loop of ribbon to the coat lining beneath the armhole and slip the scarf through the loop. If you're someone who tends to lose scarves, this is for you!

Getting water spots out of silk ties
● Rub the spotted area when dry with another part of the tie.

The sheer look
● To get a real sheer look with see-through blouses, match bra to skin color, not to the color of the blouse.

Instead of buying a new bra
● If your budget is tight and your bra is too (maybe you're pregnant or you've put on some weight), buy a bra extender in a notions shop.

To polish metal buttons
- Here's how to clean buttons without removing them or getting metal polish on fabric: Insert the prongs of a fork between the button and fabric, then clean.
- Or put a cloth around the crook of the button, hold it tightly between your thumb and forefinger and polish away.

Cleaning discolored shirt buttons
- Place a piece of plastic wrap behind the button and rub at the stain on the button with a pencil eraser.

Protect wooden buttons
- Wooden buttons may swell and crack when wet, so either remove them from a sweater before washing it or have the sweater dry cleaned.

Gown protector
- If the weather's the worst and your gown's the best, try this: Cut leg holes in the bottom of a heavy-duty, large-size trash bag, step into it, tuck your gown in and pull it up to cover to the waist. You may look strange en route, but your gown will look great when you're at the party.

Slip substitute for gowns
- Wear a long, light-colored nightgown with thin straps.

Clinging slips
- If it's clinging to your legs, turn the slip inside out and you've solved the problem.

Hints for tints
- Help disperse dye thoroughly by adding water softener to the dye or tint bath. Unless directions call for straining, this tip will work.
- Tie the dye powder in a piece of lightweight cloth, then pour boiling water over it until all the dye is dissolved. This should rid one of the problem of clumped-together, undissolved dye particles that can spot what you're dyeing.

Renewing a crease

● To renew a crease in washable knit pants, squeeze out a press cloth that has been dipped in a mixture of one third of a cup of vinegar and two thirds of a cup of water. Use the cloth to press the fabric with a hot steam iron.

Preventing perspiration rings

● Spray the garment with fabric protector on the inside and outside before wearing. No more unsightly rings!

Patching sweaters with panty hose patches

● If the elbows of your sweaters tend to become worn and threadbare, protect them by sewing a piece of nylon panty hose (or stocking) inside the elbows. The patch is soft and sheer enough not to show, but it's strong.

Stretched-out sweaters

● If the ribbing at the cuffs and waistline is stretched out, dip the ribbing — but not the rest of the sweater — in very hot water. It should shrink back to normal.

Brightening white sweaters

● This isn't always foolproof, but it's worth a try. Brighten white woolen sweaters that have gotten dingy by freezing them after they've been washed. An hour in the freezer should do it. Remove, let them thaw out, and dry as usual.

Washing leather gloves

● Use a hair shampoo with lanolin to help restore the natural oil of the skins.

Keeping gloves in shape

● After washing wool ones, insert an old-style clothespin into each finger.

White glove care
- Add a small amount of liquid starch to the rinse water when washing white gloves. Starch makes them look new and helps cut down on soiling.

Gloves vs. mittens
- Fingers stay warmer in mittens than in gloves. In mittens, warm air can move all around the fingers and there's less surface area exposed to the cold.

To remove mildew from leather bags
- Wipe them with a cloth moistened in a solution of one cup of denatured alcohol and one cup of water. Dry in an airy place. Works for mildewed shoes, too.

Waterproofing (and dirt-proofing) straw handbags
- Spray them lightly with fabric protector. Let the spray dry before using the bag.

A second life for panty hose
- When panty hose becomes nubbly but still wearable, turn it inside out. You have a new pair instantly. Works with tights, too.

Which one has the run?
- You'll know which panty hose to wear if you just cut the labels off the ones with runs. Wear the good pairs with skirts and the others under pants.

Teaching old hose new tricks
- Stuff sagging upholstery or pillows.
- Wrap around wire coat hangers to keep clothes from slipping off.
- Snip legs off at the crotch, then snip the foot off and use the leg portions to slip over a cast for arm or leg. Helps casts to slide into clothing easier.

- Stuff a water bottle and use as a kneeling pad.
- Stuff a toe with catnip, toss to the cat and make him happier today.
- Stuff with mothballs and keep in a closet or cedar chest.
- Wrap a package if you're out of twine. Snip legs off and tie around the box. Cut the knot ends short.
- Store plant bulbs in the foot and hang them high and dry.
- Use as a strainer or trap on the skimmer basket of a swimming pool.

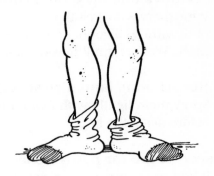

Keep 'em up
- If they're falling down, cut the elastic bands off an old pair of knee-high nylons, slide them up over the socks to the top and fold down once along the cuff.

Keeping tootsies toasty warm
- Fashionable, high-heeled boots often fit too snugly to be worn with socks, so when it's really cold, wear thin cotton anklets over panty hose. Anklets help insulate feet without adding bulk.

Where to put those dripping wet boots
- Save your floors from getting wet. Put a couple of those plastic flats you find in plant shops and stand boots on them, at the door. The flats are easy to rinse off later.

Cleaner shoelaces
- Coat eyelets of white shoes with a clear nail polish to prevent discoloration of laces.

To untie muddy or wet knots from shoes
- Use a crochet hook.

Smooth-drying canvas shoes
- After washing them, push a smooth stone into each toe. Shoes will keep their shape and the toes won't pucker.

Helping white shoes stay white
- If you rub them well with waxed paper after polishing, smudges will wipe right off with a damp piece of paper towel.

Homemade shoe deodorizers
- Cut an old sock in half at the heel. Sew or tie closed one end of the top piece to make two pouches, fill the pouches with an eighth of a cup of baking soda, then tie the pouches closed. Place one pouch in each shoe to help absorb shoe odors and keep closet smelling fresh.

Good for the sole
- New shoes are less slippery if you rub an abrasive cleanser (such as Comet or Ajax) on the sole.

Putting It Away

Storing handbags
- Use cotton flannel bags, not plastic, for leather handbags not in use. Plastic won't allow air to circulate, so leather will get dry. Stuff bags with tissues to maintain shape.
- Or use old pillowcases.

Storage in corrugated boxes

- If you live in a humid climate and plan to pack away clothing in corrugated boxes, first coat the box and cover with thinned shellac to keep out moisture.

Storage in oatmeal boxes

- Make a hatstand out of round oatmeal boxes. Decorate with construction paper or contact paper.
- While you're at it, use the inside of the box to hold gloves, scarves and other out-of-season accessories.

For keepsake

- Preserve baby's treasured christening gown by putting it in an airtight plastic bag, then pack it away in a box. It'll be like new for your baby's baby.

A cover-up

- When storing several items of clothing on hangers, bunch hangers together and cover with a large old shirt.

Homemade sachet

- A spiced tea bag can make a great-smelling sachet. Place it in a piece of nylon netting and tie with a bow.

Adding shelves

- Take advantage of space at the top of the closet by adding an extra shelf above the existing shelf to hold the overflow.
- And use those rubberized racks for stacking plates and for elevating shelf space to hold handbags upright.

Uncrowding closets

- File notches about an inch apart on a wooden clothes pole. Notches will keep clothes on hangers from being jammed too close together and wrinkling.

Stronger hangers
● Don't have enough wooden hangers to hold those heavy clothes? Use two wire hangers together and tie them with a twist tie to get the needed strength.

Pants hangers that have lost their grip
● Wind rubber bands around each end of both sides of the hangers.
● Or, put a strip of adhesive-backed foam insulation on the hangers.

The Tag End

Selling your clothes at a garage sale
● If you're sharing a sales spot, make sure each of you labels items with different color tags. As items are sold, place all color-coded tickets into one container, then tally each one up when the sale is over.

Bag it

- Instead of trying to sell lots of small items individually, make up grab bags and mark them according to category such as "sewing" or "kitchen" and price as a package.

Laying it out

- If you don't have a clothes rack, a fine substitute to display clothing is a metal ladder. Place it across two widely spaced objects that are tall enough so clothes won't trail on the ground, but not so tall that shoppers can't get to the items. Make sure the ladder is secure, then hang the sale clothes on both sides.

COOKING

Cooking from A to Z
A Is for Appetizer

When there's no time to fuss
- Stuff button mushrooms with cheese spread.
- Dot melted butter and sprinkle Parmesan cheese on canned artichoke hearts. Put under the broiler a few seconds or until cheese melts.
- Garnish smoked oysters with lemon wedges and serve on lightly buttered toast fingers.
- Add two tablespoons of mayonnaise to a container of plain yogurt. Toss with fresh herbs and use as a tangy dip for fresh cut vegetables.
- Open a jar of stuffed grape leaves, squeeze lemon on them and serve with fancy toothpicks.

Out of the freezer, onto the hors d'oeuvres tray
- Defrost cheese- or meat-filled tortellini and cook them according to the directions on the package. Toss them with a light garlic dressing, sprinkle with Parmesan and serve with toothpicks.
- Defrost a box of spinach soufflé, then spoon into large, fresh mushroom caps until filled but not overflowing. Bake in a 375° oven for about fifteen minutes. Recipe yields about two dozen.

B Is for Baked Goods

Biscuits
- To save the step of rerolling scraps of dough when using a stamping mold, try this instead: Pat the dough into a half-inch-thick square and cut it into smaller squares with a knife dipped in flour. Since dough isn't overly handled (as with stamping molds), biscuits aren't likely to be tough.

That's the way the nut bread crumbles
- Ideally, nut bread should be stored at least twenty-four hours before it's sliced. If you can't wait, it should at least be cool. Use a sharp thin-bladed knife and cut with a light sawing motion. Don't press down on the bread.

Do-it-yourself melba toast
- Cut each slice of one loaf of thin-sliced white, rye, wheat or pumpernickel bread in half. Place slices on an oven rack or cookie sheet and bake in a 200° oven for about half an hour. Turn the slices a few times so they're evenly done. Toast is done when it's nicely browned and completely dry. Store toast in a plastic bread bag for freshness.

A taste of honey
- You can overmeasure honey since it clings to the underside of the spoon. For a precise measurement, don't put the spoon directly into the jar, but pour the honey from the jar onto the measuring spoon.
- You can substitute honey for half the sugar called for in a recipe. Just be sure to also reduce the liquid by one fourth.
- Bake any cake with a substantial amount of honey at a slightly lower temperature than cakes containing only sugar.

Improving inexpensive cake mix
- Add one tablespoon of butter to the batter for a richer-tasting cake.

Preventing "holes" in angel food cake
- These holes are air pockets from where the stiff batter didn't touch the pan. Run a knife in a spiral pattern through the batter in the tube pan, pushing it against the walls of the tube for a perfect cake.

Throwaway cake plates
- Cover an old record album with aluminum foil. Works just as well and you don't have to lug it back home if you're donating a cake or pie to a cake sale.

● Or, save the bottoms from cardboard pizza boxes and use.

Sweet frosting idea
● Mix a fifteen-ounce can of crushed pineapple with three and three-quarter ounces of instant vanilla pudding. Great for topping a lemon or yellow cake.

Icing ideas for birthdays or Valentine's day
● Add a small amount of red, unsweetened powdered drink mix to powdered sugar frosting. You'll get a pink color and great flavor.
● If you use lots of maraschino cherries, save the juice and substitute for some of the liquid in your cake batter mix. Cherry juice turns the cake mix pink and adds a touch of sweetening.

This hint's write-on!
● When decorating on cake frosting, use a toothpick to trace the design or message, then use an empty mustard squeeze bottle filled with frosting to fill it in.

Filling an icing bag
● Before you fill an icing bag, place it in a tall iced tea glass, with the pointed end extending into the glass. Now when you fill the bag, the icing will go in neatly without spilling.

Prevent dough from jamming in the mixer
● If cookie dough jams the beaters and wears down the motor in your electric mixer, try using just one beater.

Rolling dough on nonslipping waxed paper
● If you roll out pastry dough between two sheets of waxed paper, dab some water under the bottom sheet and it won't skid away.

Leftover pumpkin pie filling
● Fill cupcake liners in a muffin tin with filling and bake with the pie. Add chopped nuts for a great pumpkin pudding dessert.

Heating rolls
● When the oven is too hot or full, wrap them in foil. Heat on a low setting in a crock pot.

Quick and easy doughnuts
● Use refrigerator biscuit dough. Just cut a hole in the center of each "biscuit," then fry in hot oil until browned. Drain, add toppings.

Don't go nuts making perfect doughnuts
● The more egg yolks in doughnut dough, the less grease they'll absorb when fried.
● If you let doughnuts stand about one to five minutes before frying, they'll be less greasy.
● Turn doughnuts frequently so they don't crack in the hot oil.
● A few potato slices added to the oil will keep doughnuts from burning.

Easier-to-cut meringue
- Sprinkle about one tablespoon of granulated sugar over meringue before browning.

Self-rising flour for piecrusts
- If you're out of regular flour, you can use the self-rising kind, but don't add salt. It's already got some in it. The piecrust will taste fine though it will be a bit mealy in texture and not as flaky as you are used to.

To soften dried pastry
- Place in an airtight container for twenty-four hours with a slice of fresh bread.

Adding oatmeal to piecrusts
- Sprinkle about four tablespoons of quick rolled oats on a piece of waxed paper, then roll your dough over it. Oatmeal crusts give fruit pies a nuttier flavor — and extra nutrition.

Saving your dough
- Roll leftover piecrust and cut into strips. Sprinkle with parmesan cheese and bake for an inexpensive before-dinner treat.

B Is for Beverage

Groundless coffee
- When you measure coffee directly into the basket of an electric coffeepot, place a thimble over the top of the stem so no grounds get into the pot.

Nutritious slush drinks
- Freeze orange, apple or pineapple juice in ice-cube trays. Drop the frozen cubes in a blender with enough cold juice to get it to the right consistency.

Keeping tabs on your tea bag
● Slip the paper tab that's attached to the tea bag under the cup before pouring water and the tab won't be pulled into the cup:

Tea-riffic flavorings
● Add fragrance and flavor to your tea by keeping a few pieces of dried orange rind in the canister.
● Or, make your own flavored tea by adding a drop of any flavor extract to a cup of hot tea.

Did you know?
● Dry table wine will lose a good 85 percent of its calories and all its alcohol when stirred into cooked dishes. That means a 125-calorie glass of wine will only give you 11 calories in food. Pretty good — and so's the flavor.

...And B Is for Breakfast Food

Lumpless cereal
● Prevent lumps when cooking hot cereal by starting with cold water instead of boiling water, as directed. You'll find the texture much smoother.
● Or, use a wooden spoon and pour the cereal onto the back of it while mixing it into the water.

Yogurt-flavored pancakes
- Add fruit-flavored yogurt drink (available in the yogurt section of the supermarket) to the mix. Substitute it for the liquid indicated in the recipe, then add milk to thin the batter to the desired consistency. Pancakes are yummy.

Syrup substitutes
- Add a bit of water and a dab of butter to any fruit jelly and heat in a saucepan.
- Or, use the syrup from canned fruit. Thicken it with a little cornstarch, heat and serve over pancakes.

Fluff up your waffles
- Use buttermilk and a pinch of baking soda instead of whole milk to make your waffles soft and light.
- Let the waffle cook a minute or two before closing the lid on the waffle iron. You'll get fluffier waffles.

Too many waffles?
- Freeze them. When needed place on an ovenproof plate, top with syrup and butter and heat under the broiler. Waffles are ready to eat when the syrup sizzles.

Sneaking in a little nutrition
- Add some finely chopped spinach to leftover waffle batter and have vitamin-rich waffles for lunch.

C Is for Condiments

Homemade catsup
- Combine one cup of tomato sauce or mashed or canned tomatoes with two tablespoons of vinegar, one-quarter cup of brown sugar, one-quarter teaspoon of cinnamon and a dash each of allspice and ground cloves. A delicious change from bottled catsup.

The last of the catsup

● Add a few ingredients to bottled (or homemade) catsup and make it into a tasty salad dressing. Add one table-spoon of vinegar, two tablespoons of oil and an eighth of a teaspoon or so of Italian seasoning to a nearly empty bottle of catsup. Shake well.

Dressing up dressings

● Improve bottled blue cheese dressing with one-half finely chopped hard-cooked egg and one teaspoon of dry white wine to a half cup of dressing.

● Freshen up each half cup of Green Goddess dressing with one tablespoon of chopped chives and a half to a full tea-spoon of finely chopped parsley. Add to one-half cup of dressing.

● Add flavor and crunch to a half cup of Italian dressing with two tablespoons of chopped parsley, a half clove of finely chopped garlic, two tablespoons of grated carrots and a half teaspoon of lemon juice.

The "mayo" clinic

● Use an empty, well-washed-out Worcestershire sauce bot-tle and you can add oil drop by drop when making mayon-naise. The cap design makes for perfect control with no overpouring.

● Or, use a syringe-type baster.

● Keep it on one of the least cold shelves in the refrigerator, such as on the door, to prevent it from separating.

● To reduce the risk of its separating, use a warm bowl, add egg yolks and mix in a teaspoon of mustard, then add the oil.

● Or, after making the mayonnaise, stir in a tablespoon of boiling water to keep it from separating in the refrigerator.

● If your cake recipe calls for oil and you've run out, you can substitute an equal amount of mayonnaise. You'll get a moist, delicious cake.

● If you rub the skin of a chicken with mayonnaise before baking, you'll get crisp, brown skin.

- If you're out of oil when frying meat, use a few tablespoons of mayonnaise instead.

D Is for Dairy Foods

Butter that's spreadable
- Soften it in a bowl of warm water before removing the wrapper.

Cutting cheese straight
- Keep your eye on the part you're cutting from, not on the piece that's being cut off, and you'll cut it straight every time. This works for bread, too.

Shredding your own
- Buy cheese by bulk and shred it yourself and you'll save money. A one-half pound block of Parmesan cheese yields three heaping cups of fresh cheese and a half pound of American or cheddar cheese yields two cups when shredded.

Prevent boxed cheese from sinking in the middle
- Turn cheeses like Brie and Camembert upside down to prevent sunken middles.

Saving leftover cheese
● Save the tail ends of different kinds of cheese, then grate and freeze in an airtight bag. Top dishes that call for melted cheese like veal parmigiana or use in omelets.

Beating egg whites
● Use glass or metal mixing bowls, not plastic. Plastic tends to retain grease, which can prevent whites from whipping up.

Warming eggs for beating
● Since eggs beat to a greater volume when they're at room temperature, try this if you don't have time to let eggs warm up naturally: Put them in a bowl of warm water for a few minutes, then crack and separate.

Free egg poacher
● Make one by removing top and bottom of a tuna can. Set the remaining ring in a skillet of simmering water and crack the egg into it.

Finding fresher eggs
● Buy brown eggs one week, white the next. You won't wonder which eggs are the freshest when you alternate colors.

Money saver
● If you shop in a warehouse-type grocery store, bring empty egg cartons and buy eggs from the bulk-packed cases. You'll save a little, but every bit counts.

Eggs-cellent solutions for baking
● If a cake recipe calls for two whole eggs and the frosting calls for egg whites only, save on eggs by doing a little juggling. Use one whole egg plus two egg yolks in the cake and you'll have two egg whites left for the frosting.

Devilishly delicious
● To prepare large quantities of deviled eggs quickly, put the egg yolks through a potato ricer.

Transporting deviled eggs
- Wrap them individually in plastic wrap, gathered loosely at the top and place in an egg carton. When carton's closed, eggs are kept upright and separate. And, there's no messy dish to carry home if you're picnicking.

What to do with uncooked eggs
- Grease the top of a double boiler and put the uncooked eggs into it without stirring. Cook until firm. Chop by criss-crossing with a sharp knife for instant hard-cooked eggs, chopped to perfect salad-making size. Keep this hint in mind after blowing eggs at Easter. It's an egg saver.

Condensed milk
- Mix one and one-eighth cups of dry milk powder and a half cup of warm water in a double boiler. Stir in a three-quarter cup of granulated sugar and stir over gently boiling water until sugar is dissolved. Remove mixture from double boiler and milk is ready.

Wine in dairy dishes
- Add wine first to dishes that contain cream, eggs or butter and they won't curdle.

Low-calorie topping substitute
- Beat an egg white and a sliced banana with an egg beater until stiff. Mix about four minutes or until banana is completely dissolved.
- Here's an idea: Pour a can of evaporated milk into a freezer tray and freeze until ice crystals form. Whip it as you would heavy cream.

F Is for Fish

Did you know?
- The darker the flesh of a fish, the higher it is in calories.

- Tuna labeled "chunk" is made up of irregular pieces, while "solid" is mostly one piece. There's almost no difference between them nutritionally, though solid packs sell for a higher price.

Poaching fish
- Poached fish will be firmer and whiter if you add a bit of lemon juice to the cooking liquid.

Improving the color of lobster
- Rub the shell with salad oil if you'll be serving the whole lobster cold.

Stretch the salmon
- Mix a quarter cup of chopped onion and a half packet of corn bread mix with a can of undrained salmon.

...And F Is for Fowl

Easy way to cut an uncooked chicken
- Chill thoroughly. Just short of frozen is best.

Finger-lickin' chicken
- Here's how to get a light and delicate crust on coated, fried chicken. Add in about three quarters of a teaspoon of baking powder to the batter and use club soda as the dipping liquid.

Deep-fat-frying chicken
- Choose corn or vegetable oil over sesame, peanut or olive oil when deep-fat-frying. Corn and vegetable oils have higher smoke points and don't burn off as fast.

Thawing turkey in a hurry
- Set it in an ice chest or picnic cooler placed in the tub. Fill the cooler with ice cold water and change it frequently.

Instant unstuffing
- Unstuffing a turkey is a snap if you first: Tuck a cheesecloth bag into the turkey with the open edges hanging out. Spoon dressing into the bag until it's filled, then fold the bag closed. Tuck the edges up into the cavity. When the turkey's cooked, open the stuffing bag and spoon a little out. Then grasp the outside edges of the bag firmly and pull it out of the turkey.

Making a less expensive turkey a lot juicier
- Try this: Put the bird in a covered pot on top of the stove and let it steam in about a half inch of water for about an hour before popping it into the oven for browning. Basting isn't necessary. But be sure the water doesn't steam-dry. Your bird should be as tender as the prebasted ones.

Leftover turkey idea
- Reheat on top of stove in a pan with about one-quarter cup of milk. Be sure to reheat over a low flame and turn the meat so both sides get heated evenly. Don't let the milk boil. The bird will taste fresh from the oven.

Stretching stuffing mixes
- Add two or three cups of bread crumbs or corn bread cubes per package of stuffing mix. You'll be able to serve six extra people and the taste's just about the same.

...And F Is for Frozen Food

Freezing large quantities of fish
- Here's how to get your catch from your cabin back to town: Use disposable aluminum foil loaf pans. Lay enough ready-to-cook fish for two servings in each pan, cover with water and freeze. When solidly frozen, slip block out of pan, smooth the edges of the ice block and then wrap in heavy-duty foil. Rinse loaf pans and reuse.

Ice cubes in a hurry
- Leave two or three cubes in the ice tray you're emptying. Fill remaining slots with cold water. The frozen cubes will make the others freeze faster.

Fast freezing
- Cover unfrozen food with another food that's solidly frozen and it will freeze faster.

G Is for Grains

Prevent long grain rice from boiling over
- Cook it in a tea kettle with a removable lid. The spout keeps the rice from boiling over and the rice cooks faster, too.

Different flavors for rice
- Cook with bouillon, vegetable or meat broth instead of plain water.
- Or, add unsweetened pineapple juice for a tasty "pudding" treat.

Leftover rice: makes tastier bread
- If you make your own bread, add up to one cup of rice per loaf. Bread's flavor and texture really improves!

Avoiding pasty pasta
- Allow at least five quarts of rapidly boiling water for one pound of pasta.
- And add a tablespoon of oil so the water doesn't boil over.

M Is for Meat

To freeze bacon for a small portion
- Accordion-pleat some waxed paper and insert a strip of bacon in each fold. Wrap in freezer wrap and freeze.

For soft but fully cooked bacon

- Keep a glass of water near the stove burner and dip each raw slice of bacon in the water before dropping onto the skillet.

Clarifying bacon fat

- Pour the cooled fat into a can that's got a few inches of cold water in it. Burned bacon bits will sink while the pure fat rises and hardens. Skim off fat and store in the refrigerator.

Beef marinade

- To make a marinade if you're out of wine, marinate meat in beer for a few hours.

No-bake meat loaf

- Form patties with the meat loaf mixture and brown them in a skillet in one tablespoon of hot oil. Drain off excess fat, top with one 10½-ounce size can of tomato soup and an 8-ounce can of tomato sauce. Simmer and serve over rice.

Mixing meat loaf

- If you don't like the mess of mixing meat loaf with your hands, use the dough hooks on your food processor or mixer.
- Or, use a pastry blender.
- Or, wear rubber gloves or plastic bags tied with rubber bands and work with your hands.

Browning meat in a microwave

- Brush it first with either soy sauce or Worcestershire sauce. Meat will have a better taste and appearance.

Freezing hamburger patties

- Alternate patties between waxed paper sandwich bags. This eliminates the job of separating them since the double thickness prevents sticking.

When ham's too salty
● After baking it partially, drain all the juices, then pour twelve ounces of ginger ale over it and bake until done.

Substitute for flour to coat pork chops
● Try prepared pancake or waffle mix.

P Is for Produce

Lunch in an apple
● Here's a low-cal idea. Slice off the top of an apple. Hollow out the core, leaving the bottom intact. Brush the hollow with lemon juice, then fill with chopped fruit salad or chicken salad and fit the top of the apple back on. Wrap in plastic until ready to eat.

Buying and cooking cabbage
● To avoid buying a cabbage with a strong flavor and a coarse texture, be sure no separate leaves are growing from the main stem below the head.
● A wooden pick or two inserted through the cabbage wedge will hold the leaves together during cooking.

Cooking corn
● Drop ears of corn into a basket used for french frying potatoes, then immerse them in a pot of boiling water. You won't have to fish around for the ears: Just lift out the basket when corn is done.

Dates and figs
● To remove the pits from dates, cut with a scissors dipped in water.
● To chop dates or figs, spray the chopping blade with non-stick cooking spray. Sticky fruits won't stick.
● Eat fresh figs at room temperature. Chilling tends to tone down the flavor.

Buying sweeter eggplant

● One with fewer seeds will be sweeter than one with lots. Here's how to know: Check the bottom of the eggplant for the shape of the indentation, which is about the size of a dime. If the indentation is oblong or oval, it will be loaded with seeds. Instead, buy one with the round indentation.

Lettuce leaf burritos

● Use lettuce leaves instead of tortillas for a delicious low-cal burrito. Fill with chopped vegetables.

Cooking okra

● To avoid discoloration of the pods, don't cook okra in brass, copper or iron pots.

Keeping cooked onions whole

● Punch a hole through the center of each onion with a metal skewer.
● Or, stab the peeled onion with a fine-tined fork about half-way through to the center. The onion should keep its shape.

Peppers

● The larger the pepper, the milder its flavor; the smaller, the hotter the taste.

Potatoes

● Stick three or four toothpicks into one of the long sides of the potato before placing it in a microwave oven. Stand it on the toothpicks, and it will cook more evenly.
● Use the water from boiled potatoes when making gravy.
● And, use the water from boiled potatoes instead of milk when mashing potatoes.
● If you peel more potatoes than you need, cover them with cold water and add a few drops of vinegar to the water. Refrigerate and they'll keep for several days. Be sure they're completely immersed.

- If you have leftover baked potatoes, cut them into thick slices. Deep fry them, then dip in garlic butter after they're well drained.

Squash
- Squash that's hard-shelled will have a sweeter flavor if refrigerated rather than left out until used.

Zucchini
- You won't have to hollow it out first if you steam it, then make a long deep cut down the middle. Works great for small zucchinis.

Cooking leftover vegetables
- Put them in individual cooking bags, seal and throw into a kettle that's one third full of boiling water. Cover and allow to steam cook.

S Is for Sandwiches

Did you know
- A one-pound loaf of thin-sliced white bread yields more slices than a one-pound loaf of thin-sliced whole wheat. Why? Could be the coarser texture (and heavier weight) of the whole wheat.

Bread substitute
- Make a sandwich with two waffles. Add ham, cheese and mayonnaise for a different treat.

Easy slices
- Sandwiches that include cheese are easier to cut if you put the cheese slices on the bottom.

Another cutting tip
- When cutting into homemade bread for sandwiches, chill the bread at least half an hour before slicing. It'll be easier to cut, especially if you want the thinnest possible slices.

Crunchier peanut butter
- Add some crunch to peanut butter with coarsely chopped carrot bits. It's a good way to sneak in a little more nutrition.

Softening peanut butter
- Add one teaspoon of hot water. Stir.
- Or, drop in a little sesame oil. It not only restores creaminess, but increases the nutty flavor.

...And S Is for Soup

Removing bones from homemade soup
● Put soup bones and all other ingredients that will be removed from the broth into a vegetable steamer. Just lift out the steamer when the soup's done.

Freezing soup stock
● Pour it into loaf pans, freeze, then remove from the pans and wrap in freezer paper. The blocks are easy to store and take up a minimum of freezer space.

An idea for leftover chicken soup
● Use it instead of oil to stir-fry favorite recipes.

...And S Is for Spices

Bay leaves
● Place bay leaves in a tea ball for easy removal from stews or sauces.

Dill
● Run it under cold water when you buy it and keep refrigerated in a glass jar with a tight lid.

Mint
● Sprinkle a pinch of sugar on mint when you chop it. Sugar draws out the juices and makes chopping lots easier.

Dry mustard
● A half teaspoon added to a flour mix for frying chicken adds great flavor.

Parsley
● Did you know that parsley lends freshness to other herbs? Mince equal amounts of it with dill, basil or rosemary.

Pepper
- A few peppercorns in the pepper shaker will keep the holes from clogging while giving more of a fresh ground pepper taste.

How much spice is enough
- If a recipe calls for fresh herbs and you've only got the dried varieties, use one third the amount of dried herbs as a substitute. Drying concentrates flavors.

Spicing up a salt-free diet
- Blend together two and a half teaspoons each of paprika, garlic powder and dry mustard, five teaspoons of onion powder, one-half teaspoon of ground black pepper and one-quarter teaspoon of celery seed. Keep a shaker of this mixture on the table and pass up the salt.

Rehydrating dried herbs
- If you don't like the crunchiness of dried herbs in cold dips and sauces, try this: Place the herbs in boiling water for thirty seconds, strain, pat dry, then use.

Drying herbs
- If you've grown an abundance of parsley, dill or chives, dry and store them. Wash and drain fresh herbs well, then place between paper towels in the microwave for ten minutes on medium setting. Cool herbs, then crumble or chop in a blender. Store in airtight jars.

...And S Is for Sweet Things

Breaking up chocolate
- Leave it in the wrapper, then pound with the textured end of a meat pounder. This works the best of any method.

Melting chocolate
- Here's a handy measurement to know: Two ounces of melted chocolate yields one-fourth cup. If you grate four ounces of chocolate, you'll get three quarters of a cup.
- Having trouble getting melted chocolate out of a pan? Add a little flour to whatever is left in the pan and mix well. The chocolate should come out easily to add to your cake batter.
- If you're melting a small amount, place it in one cup of an egg poacher. Use the poacher like a double boiler.
- For larger amounts, put it in a boilproof plastic bag and dip it in hot water. When the chocolate has melted, cut off a corner of the bag and squeeze it out.

Shaving chocolate for curls
- Curls are larger when made with milk chocolate than with semisweet.

Oh! Fudge!
- Fudge won't sugar if you add a dash of cream of tartar.
- Break up fudge that didn't set or crumble to use in cookies, cakes and puddings.

Finding the capacity of a mold
- When you don't know how much liquefied dessert you need to fill a decorative mold, fill it with water then pour the water into a one quart measuring cup. You've got your measurement now.

Set rules for gelatin
- Never put gelatin in the freezer to set. It'll turn gummy in appearance and the surface will crack.
- To "quick set" gelatin, try this: Pour the gelatin mixture into a large bowl. Add one cup of boiling water, stir, then continue to add ice cubes until the mixture starts to set. Remove unmelted ice, transfer to mold, and place in refrigerator.

Dessert ideas for gelatin

- Add two tablespoons of cherry gelatin to an apple pie mix. Besides helping to thicken the pie, it will add color and flavor.
- Add one cup of plain yogurt instead of one cup of cold water to a gelatin dessert. You'll get a creamier, tarter flavor.

Did you know?

- Ten miniature marshmallows equal one large one.

Soda gone flat?

- Here's how to tell. Squeeze the plastic bottle. If it gives, the soda's lost some of its fizz.

U Is for Utensils

Beer can openers

- The pointed end is an excellent tool for deveining shrimp.
- Or, use on those hard-to-open paper boxes that don't "press" open as easily as they're supposed to.

Meat saws

- Slice squash with a tough rind.

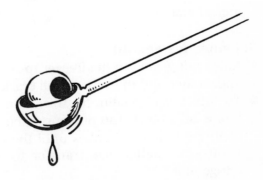

Melon ballers

- Form appetizer-sized meatballs for parties. Dip the baller in cold water occasionally to keep meat from sticking.

- Or scoop olives out of the jar without piercing them. Juices will be drained too as you lift the olives out.
- Or core pear or apple halves.

Pizza cutter
- Cut homemade bar cookies to make nice, smooth squares in half the time.
- Or score fudge.

Potato peelers
- Use the tip to pit cherries.
- Or cut orange or lemon rind peels. They'll come off without the white membrane.
- Or shred cabbage. Cut the head in quarters and hold by the core. Shred in long strokes of the peeler.

Refrigerator crisper drawer
- If you need an extra large container for salads, try a crisper drawer.

Soup ladle
- Turn it into a double boiler by placing it in boiling water and using it to melt butter or chocolate.

Z Is for...Zee Rest of the Hints in This Chapter

A clever way with coupons
- Here's how to clip coupons without destroying the paper before everyone's read it. Clip only around three sides of the coupon. The last person to read the paper can pull the coupons out for you.

Stronger shopping bags
- If you walk to the store (or send kids for heavy groceries) place the cover of a cardboard shoe box in the bottom of the shopping bag.

Marketing strategy

- List the items you need in the order you'll find them in the store. You won't keep going back and forth to the same section.
- Arrange coupons by aisles in the supermarket and save time, too. Clip marketing list to the supermarket cart with a clothespin.

Watch those purse strings

- The fastest-moving items in the supermarket are kept at eye level. Don't fall for it: Cut down on impulse buying by making a conscious effort not to pick up the first items you see. You may not need them.

Out of the bag

- When you get home, take all the groceries out of the bags and put them on a table or counter. Put all the food that needs refrigeration in one spot, then open the refrigerator door only once. You'll save time and energy.
- Instead of carrying a calculator, just round off each item purchased to the nearest dollar and put a mark on your grocery list for each dollar. You'll tally up to within a dollar or two of your total bill. Sad, but true!

ENTERTAINING

Come On to My House

Invitations with pizzazz

- Cut a picture of a celebrity or funny situation from a newspaper or magazine, paste on plain paper and write an appropriate message, followed by party details: time, place and date. Xerox and mail.
- Hand print invitations to a more formal party on a plain folding paper fan. Tie a thin silver ribbon around it and mail in a business-size envelope.
- For a party with a sports theme: Make colored paper pennants and write all the details on them in white ink. Cut the points off toothpicks and glue to the wide end of the pennant. Mail in a matching color envelope.
- Or, send beer coaster invitations. Give all information on the blank side.

Keep the party moving

- For a livelier party it's always best to have more guests than seating. This will keep your guests mingling, rather than sitting down.

- If your apartment is small, the guest list large and you only want to prepare for a party once, try this: Stagger the hours you ask guests over. The first fifteen to twenty invitations can be for 5 to 7 P.M., the next ten or so for 6:30 to 8 P.M. and the last list of people for 7:30 to 10 P.M. There'll be just enough overlapping of friends without too much of a crunch.
- Or you can have a party in the evening and brunch the next morning. Get double use of flowers, rented chairs, etc., and plan the menu to double up on ingredients: leftover ham goes into a quiche, fruit becomes fruit salad.

A house warming
- Save energy and keep everyone comfortable by lowering the thermostat before guests arrive. A roomful of people heats up quickly.

Energy saving (yours)
- Just a reminder: arrange all the foods you'll need on one shelf in the refrigerator. It's easier to remove — and you won't forget to serve any of it!

Lopsided candles
- Heat the end of the candle before inserting it in the holder and it will stand at attention.

Candlestick makers
- Clay flowerpot saucers make great, inexpensive candle holders.
- Or, use a demitasse saucer.
- Or, cover small blocks of wood with Con-Tact paper or wallpaper, then drive small nails through the bottoms to hold the candles in place.

Nifty name cards
- Drape a wide silver or gold ribbon over the center of each plate. Write each guest's name with a wide felt-tip pen or use press-on letters. Good for an anniversary or birthday party.

Pick Up Your Spirits

Champagne lost its bubbles?
● If you're serving inexpensive champagne and the carbonation fizzes out, add one raisin to the bottle. The raisin won't affect the taste, but its raw sugar will start the bubbling up again. A wine expert told me this one!

Out of ginger ale?
● Mix equal parts of Coke and 7-Up. You won't be able to tell the difference.

Hot punch
● Use a crock pot to control temperature. What could be easier?

Cold punch
● Fill a crystal punch bowl with cold tap water an hour or so before using it. Chilling protects crystal from cracking later when ice is added.

Prettier ice cubes
● Drop a rose petal into each compartment of an ice-cube tray. Fill with water, then freeze. This looks terrific floating in a drink.
● Or, try lemon peel.
● Or, strawberries.
● Or, mint leaves.
● Or, candied violets.

Extra ice-cube trays
● Use styrofoam egg cartons as trays when you need extra ice for parties.

The Gang's All Here

Keggers with a theme
- Make beer parties more interesting by varying the selection. Serve bottles of Japanese, Mexican, British, Danish and Canadian beer along with the usual American and German brands.
- Match hors d'oeuvres with two or three varieties of brew. For example, guacamole and taco chips (Mexican), bite-sized sausages (German), or batter-fried shrimp pieces (Japanese). Use your imagination.

Hot ideas for the barbecue
- You can cool down the fire without extinguishing it if you sprinkle water over it with a clean whisk broom.
- You'll know if the outdoor grill is ready to start cooking if you use this easy test. Hold your hand over the grill at the approximate level the food will be cooked — and be careful. If you can only hold your hand there to the count of three, the fire is hot. If you get to a count of five, the fire is at medium and a count of eight means the fire is low.

Picnic food for 100 guests
- 35 pounds of hamburger and 140 buns (for 4-oz. hamburgers).
- Or, 20 pounds of hot dogs (200 buns).
- And 5 gallons of baked beans.
- And 12 quarts of potato salad.
- And 20 pounds of coleslaw.
- And 50 pickles.

One-of-a-kind serving dishes
- Beef stew in a whole cooked pumpkin. Seed and rinse pumpkin with milk. Bake at 350° for forty-five minutes to an hour until tender but not too soft. Ladle in the warm stew.

- Shrimp salad in an avocado half.
- Mashed sweet potato (flavored with orange juice) in half an orange.
- Cooked and diced chicken in a zucchini boat.
- Corn salad in a whole green pepper.

Weddings with style
- Try getting space at a local botanical garden. Most of them have adjoining buildings that can be rented for receptions. And there's a bonus to marrying there: free floral "arrangements" when the garden's in bloom.
- Or, marry in the parlor or ballroom of a landmark mansion. Many of those terrific old buildings rent rooms for special occasions.

Catering the wedding
- Check to see if your caterer charges by the plate. Guests having seconds and thirds at a buffet may actually drive your total cost higher than a sit-down dinner would.

Holiday Helpers

Easter egg dye from the fridge
- You can get great colors from vegetables. Use beet juice for red dye.
- Or, cranberry juice for pink.
- Or, strong coffee for brown.
- Or, three shredded red cabbage leaves, soaked, for blue.
- Or, the skins from three medium-sized onions for tan.

The best way to dye Easter eggs
- You'll see that the dye penetrates a lot better if you wash the eggs first with a mild detergent and rinse well.
- Eggs should also be hard-boiled and chilled before dyeing for better results.

Great toss-away dye containers
- Use clean cans. Start saving the ones from soups and vegetables a few weeks before egg coloring time.

Blowing eggs
- Using a small drill bit, carefully drill a hole into each end of the egg. When the yolk's broken, blow the egg out.

Easter egg hunt
- If your kids won't eat hard-boiled eggs, organize a hunt for egg-shaped containers from panty hose filled with "grass," jelly beans and other goodies. Start saving them now.

Halloween tricks
- Cut out a jack-o'-lantern's face on a slice of very dark pumpernickel bread. Place a few slices of cheese on a second piece of bread and top with the "carved" piece. Toast the sandwich lightly under the broiler and you have pumpkin sandwiches.
- Draw pumpkin faces with a black permanent marking pen on some oranges. Stand them in a row on the fireplace mantel or windows for a cute, inexpensive decoration.

The safest, best-lit jack-o'-lantern yet
- Cut a hole in the bottom of the pumpkin just deep enough to hold a miniature flashlight. Carve the pumpkin as usual and flip on the light. It's lots safer than candles.

Christmas cards on display
- Punch a hole in the center top of each card and string a ribbon through making a knot after each card. Drape the string of cards across a mantel or kitchen window. You'll have a swag of cards to brighten a room.

Drying hand-painted ornaments
● Here's how to solve the problem of what to do with them while decorations, paint or glue dries. Stand a mug tree on a few pieces of newspaper and hang wet ornaments.

Before spraying artificial snow on windows
● Spray lightly with nonstick cooking oil like Pam. The snow will wipe right off when the holiday's over.

Trimming the tree
● Put icicles in the freezer for a few days before using. They'll be static-free.
● The baby may have outgrown her crib mobile, but it's got another life: Just detach the figures from the main post and hang them on the tree.

To catch the overflow of gifts
● Open a colorful man-sized umbrella and fill with presents. Tie red ribbon bows around the tip of each rib and a big bow at the handle. (This is a good place to keep gifts for guests who'll drop in over the holidays.)

Substitute clips for outdoor lights
● Fasten lights with spring-type clothespins.
● Or, use wrapped wire ties.

Removing tinsel from artificial Christmas trees
● Comb through the branches with a wire dog brush.

Don't toss out the popcorn and berry garlands
● Hang them outside on shrubbery as treats for the birds.
● Or, freeze popcorn garlands and use them next year. (Be sure to place in airtight containers.)

Another use for Christmas tree lights
● Use as night-lights. The darker-colored ones give a nice, subdued glow.

Protecting fragile ornaments
● Wrap them in empty bread bags, then in newspaper. The bag protects the ornaments from the newsprint. And if the ornament should break, all the pieces are contained in one bag and cleanup is easy.

Packing cookies
● Keep those plastic baskets from berries. Line each with a paper napkin and fill with Christmas cookies or candies.
● Use long plastic tomato trays. Line one with a doily, set up cookies in it, then top with a second tray. Seal with cellophane tape and wrap.
● Cover a coffee can with holiday paper, fill with candy, cookies or puzzle pieces and trinkets. Top with its plastic lid, then wrap.

Using your noggin with leftover nog
● Pour eggnog without alcohol into ice-cube trays. When almost frozen, insert wooden sticks to create "eggnogsicles" for you or the kids.
● Or, thaw frozen eggnog to use as a dessert topping for ice cream, cake or fruit. Great, even after the holiday season's over.

Happy New Year!
● Make a centerpiece of a large round clock flat in the center of the table. Surround it with noisemakers and confetti and curled paper made from leftover Christmas wrap.

Wrapping It Up

Gift Wrapping Ideas
● A matchbox makes a cute gift box. It's just the right size for earrings, rings and tie clips.
● Travel posters are the right size for a medium-sized box.

- Photocopy an eight-by-ten or an eleven-by-fourteen-inch photograph and use that for smaller packages.
- Make your own inexpensive paper with white tissue paper and food coloring. Layer three or four sheets of tissue and fold them in half, in half again and in half again. Now dip the corners only into solutions of food coloring—one color per corner. Be sure you don't soak the paper. Let the tissue dry out on newspaper, unfold, then iron flat.
- Don't toss out that bouquet of plastic flowers you're tired of. Clip the stems and use the flowers to trim gift boxes.

Ribbons from wrappings
- Instead of buying ribbons to match the paper, use a matching paper (or the same one) and cut long strips about a half inch wide. Curl with a scissor and glue down to the box top. If you don't want to use paper right off the roll, use scraps.

Bonus packaging ideas
- Pack interesting small gifts for children in a lunch box.
- Pack a necklace or other jewelry in a soft traveling jewel case.
- Pack panty hose or socks in a traveling hosiery case.
- Guest towels in a decorative tissue dispenser.

Storing gift wrap
- End the problem of torn edges and constant unrolling. Remove the paper from the tube and reroll it. Slip a rubber band around it and put it inside the tube. Paper clip one end of the paper to the tube and it'll stay in place... and in good condition.

If your gift is an appliance
- Fill out the warranty card and mail it in. Slip a card into your gift mentioning that you've done so.

Homemade gifts
● If there are any special instructions for care of what you've made (washing, cleaning, polishing, etc.), type it up and include with the gift.

All my children
● Take a group shot of your family, blow it up to poster size and have everyone write a short message over their picture to give a doting grandparent.

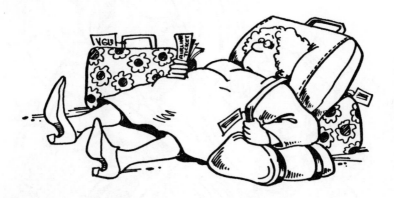

Give your holiday guests a real gift
● You may think you're doing them a favor by starting off their visit with a big night on the town. But if they've traveled a long distance, they might prefer a quiet dinner at home and a chance to relax. Give them a choice.

GARDENING
AND
FLOWERS

Some Seedy Ideas

Off to a fast start

- Place seed trays on top of the refrigerator. Extra warmth radiating from the fridge helps seeds germinate.
- Or place the trays of seeds in plastic bags, tie shut and put on top of a console TV set.
- Or when starting seeds in a window box, lay cotton balls on the soil. They retain moisture when you water. Remove the cotton when seeds start sprouting.

Soaking seeds

- If some seeds need to be soaked before planting, drop them in a wide-mouthed Thermos with warm water. Strain the seeds when they're ready.

Mixing seeds with talcum powder

- If your seeds are gathered in a clump, talcum powder will separate them so you can get better distribution while sowing.

Easter shell seedlings

- After Easter, use decorated blown eggshells for seed planters. Gently crack off the top quarter of the shell, then insert some damp potting soil. Sprinkle on a few radish or alfalfa seeds, cover them lightly with a bit more soil and keep them in a sunny window. Be sure the soil is kept moist, and in a few days, seeds will sprout.

Covering seeds with soil
- Old, large paintbrushes are great for covering seeds you've just planted in a row. Brushes give you more control and help you seed with just a few swipes.

Protecting seedlings with plastic straws
- Cut the straws into one-and-a-half-inch pieces, then slit each piece lengthwise. Slip a section around a stem before transplanting. Straws will keep cutworms from destroying plants.

Identifying plants
- Make your own markers from foam meat trays. Cut trays into triangles, write the plant's name in indelible ink on the wide end and push the pointed end into the soil.
- Or use spring-type clothespins that have the name of the plant, number of seeds and the date printed on them. Clip to the pot.

How Does
Your Garden Grow?

Rootings
- Use small glass spice bottles for rootings. The plastic pouring lids keep stems separate and delicate roots can be seen at a glance.
- When rooting any plants in a glass of water, place aluminum foil around the top of the glass, then poke holes in the foil. Insert the cuttings. Rootings will stay tangle-free and separate, and the water won't evaporate as quickly.

Feeding
- Fertilizer that has a high nitrogen content is the only type worth buying.

Fertilizing with ashes

- Once in a while, sprinkle wood ashes evenly over the soil. Ashes are about 50 to 75 percent lime and help control pH levels in soil.

Miniature greenhouses to make yourself

- Cut bottom off plastic gallon milk jugs and stand them over very young plants. Make sure to leave caps off for ventilation. Keep jugs in place until plants outgrow them.
- This one's great for vegetables: Line one side of a cardboard box with a double thickness of aluminum foil. Set growing plants in their pots inside, and cover the box with a plastic bag from the dry cleaners. Place the box on a windowsill. Since foil reflects light evenly, plants will tend to grow straight rather than lean toward the sun while the plastic keeps moisture in.

Plotting and planting

- Make even, parallel rows for planting with a child's toy wagon. Have one of your kids sit in the wagon for added weight as you pull. If the kids aren't around, use bricks.

Weeding

- Keep weeds down by laying strips of old jute-backed carpeting between rows. The carpeting won't rot and can be used year after year.

Portable chairs

- Prefer to sit while you weed? Nail a small board to an old paint can. Instant portable seat.

Scarecrows

- Here's one to make yourself: Cut a plastic trash bag into long strips and staple them to the lip of a paper cup. Glue or nail the cup to a five-foot stake and set it out in your garden. When the plastic strips blow in the wind, birds will stay away.

- Or, if you want to safeguard a larger area, string up a clothesline to cross the garden and knot the plastic strips along the line.
- Or, place pinwheels in strategic spots. Kids love them but birds don't.

Hose holder
- If your hose knocks over plants in the corners of your garden as you water, try this: drive a heavy stake in each corner. Drape the hose around the stakes toward the outside of the garden to keep the hose off the plants.

If your hose springs a leak
- Small holes are repaired by wrapping the hose with electrical tape.

Easier toting
- Keep your garden hose coiled in a bushel basket. It'll be easy to carry around. Or use the basket as a storage place.

You'll get a boot out of this

● If you'll be working in a muddy garden and you're planning to (or have to) keep running in and out of the house, try this: Set two grocery bags at the door, step into them from the garden, go into the house without tracking dirt, then step out of the bags to go back to the garden.

Make your own outdoor drinking fountain

● Install a faucet upside down.

Hauling leaves

● Use an old plastic wading pool.
● Or, bedspread.

Golf cart gardening

● You'll be using it on a different kind of green, but it works: Stash shovel, rake and pitchfork in the bag and gloves and smaller tools in the bag's pockets. Wheel the cart around with you as you work.

Attaching a trellis
- Fasten it with a hinge at the bottom and a hook-and-eye latch at the top so it can be easily pulled away from the house when you want to paint it.

Do-it-yourself trellis
- Check with companies that install chain link fences for odd-shaped discards they might be willing to give away or sell cheaply. Narrow pieces are great support for climbing plants, and they look good, too.
- Or, make a trellis with five wire hangers. First make a stake from a two-by-four cut to four feet long. Using U-shaped staples, staple the hangers to the stake, so the hooks on the hangers overlap the preceding one.

Lawn mower storage
- Before putting your lawn mower away for the season, be sure to empty out all the gas. Wrap an old towel on a wire coat hanger and insert it into the gas tank to wipe up the last bit of gas. You'll prevent any gummy residue from forming during the winter.

Protection from deicers
- Rock salt spread on driveway or sidewalks may seep into lawn or flower beds and burn your grass or perennials. Save them (and save some money) by heating gravel in your oven and spreading it on ice for traction. When the ice has melted, sweep up the gravel and use again.

Going to Pot

Repotting
- Always choose a pot that's not more than two inches larger in diameter than the old pot.
- Blooming plants should be repotted after they're done blossoming, not before.

- To help reduce the shock of repotting, give the new soil a thorough watering.
- Clay pots should be soaked in water for a few minutes before you repot your plants. This prevents the clay from absorbing moisture from the potting soil.

Drainage for pots
- Use pieces of plastic screen instead of crockery shards to cover drainage hole. Repotting's a cinch since roots grow right through the screen holes. And, a plastic screen doesn't rust.

Hang 'em high
- If you want to hang lots of plants but don't want to put in as many hooks into your ceiling, try this: Anchor one very strong hook at each end of the window and extend a decorative rod or chain between the hooks. Hang plants along the rod.

Grouping small plants
- Try a wire rack that's used for holding glassware and serving drinks. When it's time to water, take the whole rack to the sink. No need to remove the plants.

Water ways for plants
- An athlete's water bottle (a plastic bottle with a bent straw) is great for watering hard-to-reach hanging plants.

Handy plant aids
- Use a meat baster for siphoning water runoff from large pots that are too big to move to the sink.
- But don't water other plants with the water you've siphoned off in case the plant's sick.

Cup planters
- Got a favorite china cup that's chipped? Together with its matching saucer it can make a lovely planter. Several are perfect on a windowsill.

Bouquets for You

When the vase is the wrong size
- If it's too tall, crinkle a plastic bag and tie it loosely with a rubber band and stuff it into the bottom.

- Or use marbles.
- If the bouquet is small, stand a well-washed olive jar inside the vase. Filled with water and flower stems, the jar won't show in either opaque or crystal vases.
- When flowers are too short for any vase, make a center-piece by floating a plastic doily in a bowl of water, then poking the stems through the holes.

Coffee can vase
- Fill an empty coffee can with water. Cut slits in the plastic lid or punch holes in it for flowers and snap back on the can. Set the can in a decorative basket or bowl.
- Or make a needlepoint cover and slip it over the can.
- Or cover with Con-Tact paper.
- Or enamel paint.

Make cut flowers last longer

● You may have been using a pinch of salt or a crushed aspirin tablet. Now try a little liquid chlorine bleach. Like salt and aspirin, chlorine cuts down on bacterial growth in the water. Roses will stay in bloom longer, chrysanthemums will intensify in color and asters won't fade as fast with a bleach solution.

Did you know?

● Tulips continue to grow after they've been clipped, so don't be surprised if you have to switch them over to a taller vase.

Rose float

● Keep those roses that may have snapped off their stems. Just clip the stem completely off and float the rose in a shallow clear glass bowl filled with water and a quarter teaspoon of sugar.

Flower buds

● If you're buying flowers to decorate for a party, you can buy them two days in advance if the buds are closed. They'll open just in time.

● Or, hurry them open by adding warm water to the vase.

Saving money on flowers

● If you group two or three narrow-necked vases together, you'll get a dramatic arrangement with only six flowers.

● You might save some money by stopping at the florist late at night. Sometimes you can get large quantities of unsold blooms at discount. But remember, they aren't the freshest. If the guests for whom you're decorating stay past the cocktail hour, they aren't the only ones that might start drooping on the vine.

THE
HANDYPERSON

Working with Tools

Clamps for small repairs
- To hold small objects in place while glue is drying, use an old screw-on or clamp-type earring. Both are naturals.

And a clamp for larger things
- Wrap the objects in masking tape until the glue dries. It's as good as a metal clamp.

Starting nails ouchlessly
- Hold the nail between your fingers palm up — keeps knuckles out of the way.
- Or slip it between the teeth of a comb.
- Or make a notch in a cardboard scrap and position the nail in the V while you hammer.

Taking the plunge
- Putting a towel or cloth in the overflow opening when you're using a plumber's helper in sink or tub will improve the suction.

Pliers
- For a comfortable and secure grip, place a piece of rubber hose over the handles.

Screwdrivers
- For more leverage and power, the longer the screwdriver, the better.

Whetstone flower pot
- Sharpen a knife on the rim of a clay flower pot to razor sharpness. Sharpen by keeping the blade of the knife almost flat on the rim.

Sanding disk holder
● Cut a paper plate in half, stack, then tack the halves to the shop wall for convenient storage.

Storing saw blades
● Slit a piece of garden hose and insert the teeth into it.

Keep handy
● Attach several cup hooks to the top of a wooden ladder for rags and you'll have them handy as you need them.

Fix-its

Sticky situations
● If you lose the cap from glue bottles after the tip's been cut off, you can still keep the glue from drying out. Just put the cap from a ballpoint pen on the open tip.

Glue substitute
● Try clear nail polish for small repairs.

Easy drawer pull
● If a drawer's lost its pulls, open it with a suction cup or plumber's rubber plunger.

Straightening wire
● Remove the kinks and bends by passing it in and out of the tines of a fork, pulling it to straighten.
● Or, holding tightly to one end, pull as you press with a cold iron.

Protecting decorative nailheads
● When hammering in decorative nailheads, protect the finish by covering the head of the hammer with an adhesive bandage.

Stop seeing double
- If you're getting a shadow image on your TV, try wrapping aluminum foil around the rabbit ears before you rush to call the fix-it shop.

Working with dowels
- Heat wooden dowels first in a medium-warm oven to get all the moisture out before using them. When you do use them, the wood will reabsorb moisture from the air and .expand, making a much tighter joint.

Sanding small items
- It's lots easier to sand small items if you tack sandpaper to a flat surface, then rub the object over it rather than hold the object, then sand. Saves fingers.

Sanding places hard to get at
- Cut a piece of sandpaper to fit an ice-cream stick and glue it on.
- Or, use an emery board.

Covering a small crack in a vase
- Use crayon that matches. Here's how: Hold a match under the pointed end of the crayon and drip wax into the crack. Scrape away excess wax when it's cooled.

Thaw frozen pipes
- Let them thaw slowly or they could burst. Dip rags in hot water and wrap them around the pipes. Repeat until thawed.
- Or, try placing a one-hundred-watt bulb near them.

End marred walls
- Keep doorknobs from hitting and scuffing your walls by placing a small thumbtack on the spot where the knob hits.
- Or, stick a small adhesive-backed felt pad to the center of the knob.

Emergency doorstop
- Use a tablespoon turned upside down with the handle pushed under the opened door. Works best on carpeted floors.

Stop static electricity
- If you're fed up with getting those little shocks every time you open a door in cold weather, cover the doorknobs with little knitted coasters.
- If you've no time to knit, cover it with a sock or mitten.

Repairing bulletin boards
- This is great for a light-colored cork bulletin board that has become worn with age: First sand lightly with very fine sandpaper, then spray it with clear furniture polish. Good as new!

Touch-ups for linoleum
- Use automobile touch-up paint to cover scratches and burns. The paint won't wash off, even after many scrubbings.

Finishing plywood shelves
- For an interesting edging on unfinished plywood shelves, press on colorful vinyl tape instead of wooden moldings.

Make Con-Tact paper easy to work with
- Store it in the freezer first and the paper will be easier to handle and cut.
- If your Con-Tact starts to bubble, cut a small "x" in the bubble with a razor blade. Then carefully smooth the "x" into place.
- If you're applying paper in drawers and on shelves, use a blackboard eraser to make creases and bubbles disappear in a jiffy.

Refinishing cedar closets
- If your cedar closet no longer smells like one, lightly sand its surfaces. Sanding opens the wood's pores and restores cedar odor.

Smooth ways with shower curtains
- Shower curtains will slide easily if you apply a coat of petroleum jelly to the rod, then rub off most of the jelly with a paper towel.
- And to keep curtains from flapping around and tangling with your knees when the shower's on, attach magnets to the corners.

For hanging pictures
- Fishing line is as strong as twisted wire. But if it shows above the top of the frame, it's nearly invisible.
- If a picture is vibrating because it's hung near an air conditioner or other appliance, attach a piece of heavy-duty double-stick tape to the back and anchor it firmly against the wall.
- If a picture tilts despite everything you've done, try gluing a piece of wide rubber band to each lower corner on the back of the frame.
- Or, stick a blob of florist's clay on the back of the frame and press firmly against the wall.
- Have you ever thought of Velcro?

Preventing frame from marring walls
- Push rubberheaded nails into the lower corners. Buy these nails in the plumbing section of the hardware store.

Holding free-standing frames steady
- Put a narrow strip of double-faced masking tape on the bottom of the frame.

Nonglare glass

● When framing pictures, make your own inexpensive non-glare glass by covering the regular glass in the picture frame with clear Con-Tact paper.

Nonslip bookends

● Keep bookends from slipping out of place by placing a sponge-type fabric softener sheet under each bookend.
● Or, cut a rubber stair tread to fit the bottom of the bookend and glue in place.
● Or, use double-faced tape.

Make your own bookends

● Wrap two bricks in matching quilted fabric. Coordinate them if you wish with wallpaper or other accessories.

Lights On

To store a long extension cord

● Straighten out a wire coat hanger, keeping the hook at one end. Bend the hanger up toward the hook to form a large loop on which to wind the cord. Then hang the unit in a closet.

On the beam

● To prevent flashlight batteries from corroding, place aluminum foil between the spring and the end of the cap.

Power failure

● Paint the flashlight handle with luminous paint. It'll be easier to locate at night in an emergency.
● And, if you're using candles during a power outage, put the candle and holder in a large, shallow dish filled with water. Water reflects the light and makes things a bit brighter.

Being energy wise with your fireplace
- If it's more than twenty degrees outside, fire will add warmth to the room. But, if the temperature drops below twenty degrees, you may create greater heat loss by keeping the flue open for the fire.

Saving energy with outdoor switches
- A dab of bright nail polish applied to the "on" position can show you whether a light's still burning even in daylight.

Finding your keys
- Attach a piece of fluorescent tape to both sides of your house key and car keys. At night, you'll be able to fish them from your purse or pocket in the dark.
- If you constantly are locking yourself out, start carrying two keys: one on a ring, another in your change purse.

For Furnishings

Measuring for rugs
- If you're not sure how big a rug to buy, lay newspapers out on the floor to cover the amount of space you feel looks best. Measure, then find a rug to fit.

Carpet buys
- Carpets that are either very light or dark in color, with a velvet finish and deep pile, show lint and spots the most and are more difficult to keep clean. Tweed patterns and twist fiber carpeting are more practical for easy care and long-term use.

Get a grip on it
- If your throw rugs are starting to slip and slide, apply some nonskid bathtub appliqués to the bottom.

Mending upholstery

- Tears can be mended invisibly by placing fabric tape on the reverse side of the upholstery, shiny side of the tape face down. Press it with a hot iron.

Slipcovering a sponge rubber cushion

- The cushion usually sticks to the fabric, right? It won't if you cut a large plastic bag open along both sides. Slip the closed end of the bag over the end of the cushion. The slipcover should slide easily over the plastic bag. To remove the bag, just push one side of it down under the cushion, and pull it out on the other side.

Cutting foam rubber

- This can be one of the most frustrating jobs, but here's how to make it simple. Measure and draw a line on the foam rubber, put a board along the line, rest your weight on it, and cut along the line with a sharp knife dipped in water.

Moving heavy furniture

- Lift one end at a time and slide a skateboard under each end, then wheel it away.

Pulling the right cord

- Stop guessing which cord opens the pull drapes. Wrap a half-inch piece of clear tape about eye level around the right cord. It won't interfere with the pulleys.

Blocking the view

- For temporary bathroom privacy, frost the window by brushing on a mixture of four tablespoons of Epsom salts and a half-pint of stale beer.
- Or, block the view but let in light by using aquarium paint from a hobby shop.
- Or, add a few drops of dishwashing liquid into watercolors and paint on.

Keeping cupboard doors shut

- If cupboard doors keep flying open, clip a few pieces of magnetic tape and stick them on the inside of doors at the frame. They'll work like a magnetic catch.

Disguising curtain rods and window gates

- Take the rods down, wash them off and paint them as near the color of the window molding as possible, being careful not to get too much paint on the cords of the traverse rods.
- Window gates may keep you safe, but they're not the greatest-looking additions to a room. So paint them white so they look like latticework in a garden. Place pots of morning glories or other climbing plants on the windowsill and train them to grow around the gates.
- Or, hang cafe curtains on the bottom half of the window.

Taking Care of Business

Address file box
● Instead of updating address books, use a filing card system. In addition to addresses and phone numbers, use the cards to keep track of children's names, birthdays, anniversaries, etc.

Zip code finder
● Tear out the zip code map in last year's phone book and file it in your stationery box. It's handy and easy reference when you need it.
● Store last year's directory in your car. It will come in handy to look up an address you may have forgotten.

Tracking down purchases
● When enclosing a check for a mail order purchase, write the company's name and address on the check. The canceled check will be a complete record in case you want to reorder or replace a part.

Pay as you go
● Here's a great way to save money on those large semi-annual bills like car insurance. Figure out what the payment would be on a monthly basis. Deposit the prorated amount from one paycheck per month in the bank. At the end of six months you'll have the money for the payment plus interest earned.

Check this out
● Do you write checks before entering them in the ledger? You're less likely to forget to record them if you reverse the procedure and record them first.

Make a stamp dispenser

● Cut a slit in the lid of an aspirin box with a single-edged razor or mat knife. Drop a roll of stamps in the box, then dispense them one by one through the top slit.

This is a wrap

● Before wrapping a package for mailing, write the name and address on a slip of paper and place it inside the box. Should something happen to the address on the outside, the post office can get the address from the inside and send it along.

The problem is licked

● Lick the envelope where the stamp will be placed. You won't have to taste the glue and the stamp will stick on better.
● Or, rinse out an empty roll-on deodorant bottle, fill it with water and keep it in the desk drawer.

Reopening an envelope

● This isn't always foolproof, but if you seal an envelope and then remember you forgot to include something, you may be able to open the envelope by rubbing the sealed part over a hot light bulb.
● Or, hold it over the spout of a boiling teakettle.

Recycling return envelopes

● Keep those return envelopes that come with junk mail. Just cover the address with name labels or paint it over with liquid paper, then readdress.

Preventing dog-ears on legal pads

● Have you ever used a legal pad that didn't get ragged or dog-eared along the bottom? Here's how to keep pads like new: Turn the pads around. Use them with the bound edge toward you and the open ends at the top.

Another use for typewriter correction tape
● Correct spelling mistakes in handwritten letters without completely rewriting them. Just place a piece of typewriter correction tape over the errors and rub with a ballpoint pen. Most correction tape is white, but there are sheets you can buy in pastel colors that match stationery.

Renewing worn erasers
● If pencil erasers become too smooth to do their job, file with an emery board and they'll be like new.

Emergency pencil sharpener
● Use a potato peeler.

Can't find a paper clip?
● Use a long bobby pin. It's great for holding papers together snugly.
● At the piano keep sheet music pages from flipping over while you're playing. Hold them in place with a hair clip.

To find the starting end of the cellophane tape
- This is great for masking tape, too: Tuck the cut end under about one quarter of an inch. The tape makes its own little tab.
- Or, attach a penny to the end.
- Or, place a paper clip there.

File 'em away
- A multipocket shoe bag hung in the closet is a handy way to file and store bills. Mark each pocket with the category of each monthly bill you receive.

File boxes
- Make your own from large empty detergent boxes. Just cut off the top, front to back, at an angle for easy access, and cover with Con-Tact paper. Great for magazines.

A real page turner
- If you don't have a rubber finger, just twist a rubber band around your index finger. But don't twist it too tight.
- Or, use the eraser on the end of your pencil.

Don't forget!
- When you have an important task or appointment to remember, wear your watch on the other wrist. It will feel funny and be a constant reminder to get where you're going.
- If you tend to lose lists, write them out, then insert in a luggage tag with a see-through window. Hang the tag on your key chain and you won't forget.

LAUNDRY

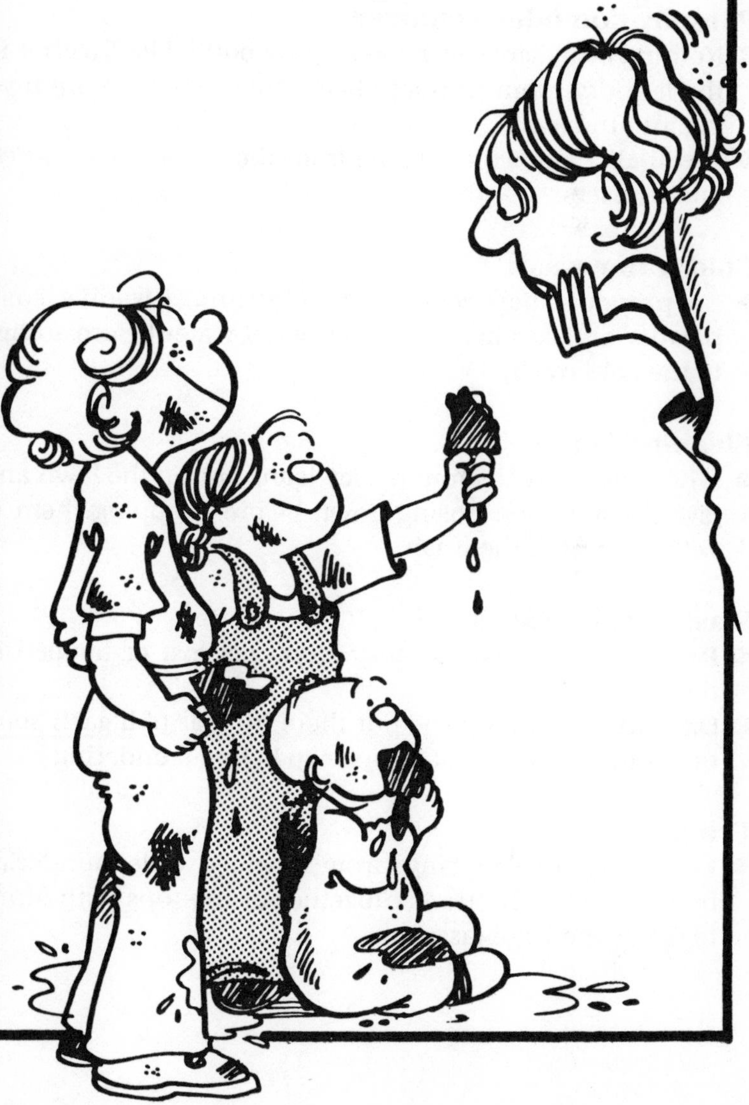

Get Ready, Get Set

Paint remover
- Use Pine-Sol or any pine-based cleaner on clothes spotted with dried-on latex enamel paint. Apply it directly to each spot, then wash.

Perspiration odor remover
- Nothing beats vinegar. Keep a spray bottle filled with it in the laundry room to treat perspiration odors before tossing into the wash.
- It will also remove the shine from the seats of dark skirts and trousers.

Cold water wash
- To prevent soap residue on clothing, dissolve soap powder or flakes in a container of hot water before adding to the cold wash.

Cleaning jeans
- When they're really muddy, lay jeans out on the lawn and give them a good hosing down before dropping them in with the rest of the wash.

Washing shoelaces
- Pin them to a towel so they won't get lost or tangled in the washer.
- Or, if they're extra dirty, put them in a jar of bleach solution and let them soak a day or so before laundering.

Fringe
- Keep long-tasseled fringe from tangling while laundering by tying each six strands together at the tops with string to make one large tassel.

Delicate baby dresses
- Collect ribbon-trimmed dresses in a mesh bag to wash with other delicate garments and use a mild detergent.

Stain reminder
- Tie a knot in a sleeve or pant leg to remind you that it needs special washing care.

Presoak pails
- Keep a large, lidded pail filled with a presoak solution in the laundry room. When clothes become extremely soiled, drop them in until there's enough for a full load of wash.

Hold That Line

Dealing with your hangups
- Wash your clothesline: Drop it into a pillowcase, tie the opening with a rubber band or string and wash in your machine.
- Blankets and tablecloths: String the cardboard tubes from paper towels on your clothesline, then hang them over the tubes to prevent line marks.
- Large shag rugs: Hang them wrong side out so the tufted surfaces will rub against each other and raise the pile.
- Sheets: hang them over two parallel clotheslines. They'll dry faster and flap less, too, and you'll save wear and tear on the fiber.
- Sheer curtains: If you're hanging sheer curtains or drapes outdoors on a windy day, slide on a clothespin every foot. or so. It will weight them down.
- Small items like socks, underwear, mittens: Save space by drying them on a four-tiered hanger hung on the line.

Did you know?
- You can ruin your clothing by letting it freeze on the line. Be sure it's thawed out before you bend it, or fibers in the clothing may break.

Quicker line drying indoors
- Place a portable fan on a table about three feet from the clothing and turn it on high. Direct the air flow from either end of the line down the center of the clothes.

Dried and True

Double duty from fabric softener sheets
- Use them once in the dryer and again in the wash.

Cutting back on softener
- You might not need as much as you're using. Try wetting your hands with water while folding laundry with static cling. Socks and lingerie will be instantly static-free.

Softener stand-in
- Pour a small quantity of liquid fabric softener on a white sock. Rub it in well, then toss the sock into the dryer with your clothes. Laundry will smell just as fresh at a fraction of the cost of sheets.

Faster drying of hand washables
- Throw a clean bath towel into the dryer with the clothes. The towel absorbs moisture and clothes will dry faster.

Drying clothes with large buttons
- Before putting clothes into the dryer, turn them inside out and button them. The buttons won't be damaged by banging against the sides of the dryer.

No-clang buckles in the dryer
● To keep overall buckles from clanging in the dryer, put each buckle into one of the back pockets and pin in place.

Speed up drying sweaters
● String up a hammock in the laundry room during the winter and use it to dry sweaters and other items. The large, open weave allows better air circulation than a flat, solid surface.
● Place the washed sweater on top of your dryer on a towel while the other laundry is drying. Heat from the dryer helps your sweater dry faster.

To get rid of excess lint on new chenille
● Tumble dry chenille spreads, draperies or throw rugs for about five to ten minutes.

Taking the wrinkles out
● New spreads, tablecloths and drapes: Put them in the dryer with a large damp towel. Set the dryer on warm and let them tumble for about ten minutes.
● New plastic shower curtains and plastic cloths: Toss in the dryer with a large damp towel, but set the dial for air fluff.

Cleaning the lint trap
● Keep a plastic mesh pot scrubber near the dryer. When trap needs to be cleaned, one quick swipe with the scrubber does the trick.

While the Iron Is Hot

More hangups
● If your basement laundry has no place to collect clothes for ironing, your problem's solved if you have a low ceiling. Mount a towel bar on one of the ceiling joists. (Use it also to hang clothes that need mending or must drip dry.)

Saving time on starching
● If you've got a lot of laundry that needs starching, spray the starch while it's wet and hanging on the line. This way, you can spray from both sides. It's ready to iron.

Starching dark cotton
● Spray on the wrong side of a dark article to prevent marks and shine when ironing.

Starching corduroy
● Turn frequently washed corduroy pants inside out and spray with starch, then iron. They'll have more body.

Starching rag or braided rugs
● Heavy starching will keep them lying flat on floors, not curling up.

To clean starch stickiness off irons
● Run iron across a piece of aluminum foil or paper sprinkled with salt. (This tip is *not* for Teflon-coated irons.)
● Or, remove the stickiness with a little rubbing alcohol.

- Or, try heating the iron, then run it over a soft, small rag. When the iron's cooled, make a paste of mild scouring powder and a little water to remove any remaining residue.
- Or, if starch sticks to the soleplate or bottom of the iron, let the iron cool and apply paste silver polish. Then wipe with a damp cloth and dry.

Filling steam irons
- Use a clean squeeze bottle dispenser from catsup or mustard and keep it near the ironing board to save steps.

Warm water sprinkling
- Clothes dampen faster and more evenly with warm water. They're also ready to iron sooner because warm water penetrates the fabric quicker.

Straight sheers
- If pressing doesn't get sheer curtains to hang straight, slip a curtain rod through the bottom hem and let them hang for a few days. When rods are removed, curtains will fall right.

Pressing cloths

- Cheesecloth is the ideal pressing cloth. It leaves no lint and retains little moisture so clothes don't get too wet.
- Or, use lightweight muslin. It works equally well.

Saving iron-on decals

- Those decals will last longer if you soak the clothes they're on in cold salt water, then press before wearing.

Ironing board cover

- Repair worn spots and tears on your ironing board: Cover with iron-on patches.
- If your cover needs laundering, remember it will fit tightly and smoothly after washing if you fasten it on the board while it's still damp.

An ironing board has other uses

- Use as a buffet table indoors and out. Just drape with a tablecloth.
- Or as an adjustable-height table for the person sick in bed.
- Or as extra desk space. Adjust it to the height of a desk and use it as an extension if you've got tons of paperwork to do.
- Or securely prop up one end and use it as a slant board when you want to relax a few minutes.

Sorting It Out

Cutting down on sorting time

- Hang clothes on the line to be ironed first and follow with clothes that need only to be folded.

Do it yourself!

- Get all family members a different color laundry basket and they can be responsible for sorting and putting their laundry away.

To each his jeans

- Can't tell those jeans apart when sorting the laundry? Use laundry marking pens and place each child's first initial on the inside pocket.

Sheet strategies

- When buying sheets for different beds, choose different colors for each one and you'll always know which sheets go where.

PAINTING AND WALLPAPERING

Painter's Pointers

Eliminating drippy paint stripper
● Put tin cans under each leg to catch the stripper as it drips down. You not only protect the floor, but you can reuse the excess stripper.

Removing old paint from ornate turnings
● Dip twine into paint remover, hold by both ends and pull back and forth between crevices.

Preventing messy paint cans
● Put a heavy rubber band lengthwise around the paint can so the band divides the can in half and makes a "partition." Use the band to wipe excess paint off the brush. Sides and top of can stay nice and clean.

Ladling paint to prevent spatters
● Use a metal soup ladle to transfer latex paint from the can to the paint tray.

Shower curtain drop cloth
● Save old shower curtains. They're heavier and more durable than many plastic drop cloths that you buy.

Trying out a large paint sample
● If you have a hard time visualizing how a color will look on your walls, buy a small amount and paint a piece of sixteen-by-twenty-inch poster board with it. The width of the board is big enough to show how a color will look on larger areas. Move it around to various spots in the room and see how it looks under different kinds of light.

Painting a stripe
- Put two pieces of masking tape on whatever you're painting — they should be parallel and placed at whatever width you want the stripe to be — then paint between them.

Slippery basement stairs
- Mix a little sand with the paint so steps will be a lot less slippery.

Painting picture frames
- Place a long, narrow stick, like a yardstick, along the center back of the frame, and lightly tack it to the top and bottom. The stick should be at least a foot longer than the frame so you can hold the frame on it as if it were a paddle. Hold the frame by its "handle" to paint. When the frame is dry, pull off the handle. There'll be no damage to the frame.

Brighten dingy closets
- Paint with high-gloss white enamel.

Kitchen and bathroom cabinets
- Finish the insides, tops, bottoms and sides before starting on the front frames and doors, so you won't smear them or get paint all over your arms.

Drawer knobs
- Remove them from the dresser and set them screw-end down into empty soda bottles. Spray paint or paint with a brush. No missed spots or painted fingers.

Radiators
- When painting with enamel paint, be sure the radiators are warm so the finish will last longer. Warmth tends to bake paint on the metal. (Check to see that the paint is specified for use with radiators.)

Repainting a refrigerator
● Before repainting a refrigerator, sand it down first with fine sandpaper, then paint with epoxy enamel or high-gloss enamel. For a super finish, use two coats of paint and sand lightly between coats.

Camouflaging hairline cracks on white walls
● Make a runny paste with equal parts of table salt and laundry starch powder mixed with water. Apply paste with a small artist's brush.

Substitute for a plaster scraper
● Use a beer can opener to scrape away broken plaster from cracks before spackling.

Substitute for spackle
● Mix white glue with finely shredded white facial tissue. Alternately add glue and tissues while kneading the mixture until it's puttylike in consistency.

Unclogging removable spray-paint nozzles
● Keep a tiny jar of paint thinner handy and drop each nozzle in after painting. Thinner stops the paint from clogging the tiny holes so they'll be clean and ready to use when you need to touch up something.

Cleaning Up

Prevent paint spatter
● When replacing the lid on a can of paint, drape a cloth over the lid, then hammer the lid back on.

Fast cleanups
● Line your roller tray with a plastic garment bag from the cleaner before pouring paint in. When finished, just lift off the bag and toss it out.

Safe paint storage
● Don't store paint in a utility room with a working furnace or water heater. Temperatures could rise too high, especially in summertime, and possibly ignite the vapors from any paint to which solvent has been added.

Ouchless stain remover
● Remove varnish stains from hands with an application of Spray 'n Wash. Spray it on, rub hands, then wash with soap and water. In most cases, it'll work better than turpentine, and it doesn't burn.

Up Against the Wall

Corking walls before papering
● Smooth out badly cracked or bumpy walls before papering by applying rolled cork over the rough walls. Hang cork with adhesive, then wallpaper right over it.

Tools at hand
● Wear a carpenter's apron when wallpapering. You'll have all tools at hand so you won't have to run up and down the ladder retrieving them.

128

Small wallpaper patterns are better buys
- Since you want to match the pattern as you hang the paper, remember that a pattern that's repeated every eight inches means a lot less waste per roll than a pattern that's repeated every twenty-four inches.

No-lump paste
- Stir first with a wire whisk. Gets rid of bubbles, too.

Tinted paste
- Add a bit of food coloring to wallpaper paste. It helps you see whether you've covered the paper to the very edge.

Paste substitutes for emergencies
- Use an egg white when wallpaper starts to pull away at the corners.
- Or, try gluing it back in place with some toothpaste.
- Or, pull the paper away from the wall as far as you can without tearing it. Brush an even, light coat of white household glue on the wall and on the paper and press it back onto the wall. Sponge off any excess glue, then run your fingers over the glued area for a minute or two.

Papering over wall anchors
- Insert toothpicks in the holes left by screws. As you paper these sections, force the toothpick points through the paper.

Wallpapering behind the radiator
- If the radiator's a few inches from the wall, this will work great. Smooth the paper down as far as you can with the brush. Wrap a hand towel around a stick and poke it down behind the radiator, smoothing as you go.

Correcting an uneven edge
- If your wallpaper is uneven near the ceiling, cover with matching border paper.

Hanging burlap on walls
- Apply paste only to the walls, never the burlap, then press on. Lap over the seams by at least half an inch.

More wall coverings
- Road maps or nautical maps
- Magazine covers
- Wine labels
- Puzzles
- Old movie stills
- Posters
- Carpet squares
- Menus
- And, for the very ambitious, matchbook covers!

SEWING
AND
CRAFTS

You Can Be Sew Clever

Finding the right fabric
- Stand at a mirror and drape the fabric against you to see if the color's right and if it hangs right for what you're making.

New sources for fabric
- If nothing tempts you at the yard goods store, check the fabrics in the home furnishings section of department stores.
- Or check the linen department.
- Or ask at upholstery shops for remnants that may sell for discount prices.

Keep it light
- Fabrics that have all-over patterns or are dark minimize sewing mistakes, while light-colored fabrics show them up.

How much fabric to buy?
- To figure out how much yardage you'll need, fold a sheet so it's the same width as the fabric you plan on using. Lay the pattern pieces out on it and measure.

Before you cut

● Be sure all pieces with nap (or patterns) are going in the same direction before cutting.

Watch out for the table

● Cover it with a plastic tablecloth to protect it from scissors scratches. Since the scissors glide over the plastic, cutting will be easier too.

Don't confuse back and front

● If you cut front pattern pieces with a plain scissors and the back with a pinking shears, it'll be easier to match the pieces.

Slip sliding away?

● To sew hard-to-handle fabrics like nylon tricot or crepe, put a piece of brown paper bag or notebook paper under them and sew through both thicknesses. Paper keeps the machine from skipping stitches and it's torn off easily after sewing.

● And, delicate fabrics won't slip off your lap while you're working on them if you wear a terry cloth apron.

No time-wasting basting

● Mark the hemlines with tailor's chalk, then use hair clips to hold the hem fold in place. Now you don't have to pin or baste at all, and there will be no pinhole in the fabric, either.

Sharper creases in pants

● Press creases into each leg before sewing the pant leg together.

Solving a pressing matter

● If you're making something that requires pressing after sewing each piece, make life simpler by putting your ironing board at right angles to your sewing machine. Adjust the board to your sitting height and you'll save time and energy.

For absolutely even stitches

● Run the fabric through a threadless sewing machine. Hand stitch by following the tiny puncture holes.

Smoother stitching

● Pull the thread over a piece of wax so the thread can slip through the material lots easier.

Fewer tangles with double thread

● Put a separate knot at the end of each thread.
● Or run thread over a sheet of fabric softener.

Rewinding a bobbin
● Mark the thread with a contrasting color crayon a few yards after starting to wind it up. You'll know you're almost out of thread when the crayon mark comes up.

Never run out of the right color thread
● Keep a few spools of clear nylon thread on hand.

Recycling bobbin thread
● Instead of unwinding it and tossing out the thread so you can rewind the bobbin with another thread, wind the last bit of thread on an empty spool and use for hand sewing.

Bobbin storage
● Use empty pill bottles.

Placing buttons
● Use a dot of white glue to hold each button in place before sewing it on. The glue washes out of most fabrics easily. (To be safe, test glue on the reverse side first.)

Reinforcing fabric under buttons

● Place a small square of fabric between the facing and garment to reinforce the area, especially if there's a small hole where the button came off.

Sewing buttons on heavy clothing

● Use six-pound or ten-pound fishing line instead of thread.

Repairing shank buttons

● If the shank of decorative buttons breaks off, sew a small flat button to the fabric, then glue the decorative button to it.

Who's got the button?

● Sew an extra matching button on the inside front of a jacket flap or the seam of a hem. It'll save you the time of hunting for one if a button falls off.

Making buttonholes

● Mark the opening of handmade buttonholes with a thin coat of colorless nail polish. When the polish dries, snip the buttonhole open and get straight, nonraveling edges that are ready to stitch over.

● When making buttonholes on a blouse, start at the bottom. If you make a mistake, it can be tucked into pants or skirts.

Buttonholing on heavy fabrics

● To snip open buttonholes on heavy fabric so they're straight, put a bar of soap under the buttonhole, and cut through with a single-edge razor blade.

● Use single strands of embroidery floss instead of thread when hand stitching them.

The Notion Department

Scissors tips
- Those rubber tips used to cover the points of knitting needles are great to protect points of small scissors. No more worry about jabbing your hand on them in the sewing basket.
- Or, you can store them in an eyeglass case.

Tweeze easy
- Keep a tweezer in your sewing basket. They're great for pulling a needle through thick fabric.
- Or, use them as aids when threading needles.

Sharp ideas for pincushions
- To make your own: Stuff a fabric square with one or two steel wool balls and sew closed. Your homemade pincushion holds pins and sharpens them, too.

Temporary pincushion
- Tuck a clean cigarette filter into the hole of a spool of thread.

Storage in your sewing basket
- Snaps: Snap them together through a piece of cheesecloth.
- Sequins: Tiny plastic boxes from breath mints are perfect.
- Trimming: Wind remnants of bias binding, seam binding and lace around large-size empty thread spools and fasten ends down with cellophane tape.

Storage boxes
- Ask at local shoe stores for extra-small boxes from baby shoes. They're terrific as sewing drawer organizers.

Sewing machine cleaning
- A clean mascara brush is just right to clean around the bobbin case and other crevices and crannies on the machine.

On the Mend

Securing drawstrings
- Tack the drawstring at the back center seam. String won't be swallowed up either one way or the other.
- Or, sew "balls" from a piece of ball fringe to the ends of drawstrings. That should keep them from slipping into the casing.
- Or, sew flat buttons to each end of the drawstring.

Sewing on appliqués
- If you dip them first into a weak solution of cold starch, then press in place, they won't slip around as you sew them on. Starching keeps them from wrinkling, too.

Double-stitching sleeves
- Try to double stitch sleeves into blouses and jackets you make to beat wear and tear on seams.

Bibs from remnants
- When making dresses or tops for tots, use the scraps to sew up matching bibs. Trim with bias tape or lace to match.

Tagging clothes that need repair
- Use the small flat closure tags that come with bags of bread and mark "fix" on them. Slip one on a hanger or over a button on anything that needs mending.

Repairs for kids' clothes
- If a favorite dress is now too short, edge the hem with a wide rickrack.

- Or, edge the hem and sleeves with contrasting bands of fabric.

Hemming ways

- When hemming a slippery or delicate fabric, make a knot in the thread every few inches. If the hem rips, you'll only have repairs between knots.

Iron-on patching jeans

- Attach patches inside the knees of kids' jeans when brand-new and they won't wear out as quickly.
- Trace around cookie cutters for fun-shaped patches.
- If a worn-out patch needs replacing, remove it by pressing it with a hot iron. The patch should peel right off.

Surprise patch

- If jeans have a hole in the knee, patch them with a back pocket from an outgrown pair of jeans. The edges are already turned under for easy sewing.

Lengthening sleeves

- When the sleeves of a great find from a garage sale are too short (a jacket or coat), attach a strip of "fun fur" to blend or match.

Fixing towels and washcloths

- Salvage worn towels by sewing over the frayed edges with decorative stitchery. Use sewing machine attachment cams for easy working.

Mending umbrellas

- If an umbrella rib breaks in a strong wind, try this mending idea: Cut a one-quarter-inch piece of wire from a hanger and use it as a splint on the broken rib. Secure it in place with strong tape.

Shower curtain covers
● Fold curtain in half and stitch up the sides to make pull-over covers for collapsible outdoor furniture.

Purse-size sewing kit
● Notch the sides of an empty matchbook cover and wind thread around it, one color per notch. Insert several needles, pointed end down, into the cardboard match striker.

Needle Little Knitting Help?

No more yarn tangles
● Make your own yarn holder with a liter-size plastic soda bottle. Carefully remove the plastic piece that fits on the bottom of the bottle. Cut the bottom off the bottle, insert a pull skein of yarn, then thread it up through the top. Snap the plastic bottom back on.
● Or when working with several colors of yarn at once, thread each through a different drinking straw to keep ends from getting tangled.
● Or, pry the top off a cleaner can like Comet, then wash and dry top off. Pull the end of each color yarn through one hole in the top.

Pay-less knitting patterns
- Borrow hardcover pattern books from your local library.
- Look for very inexpensive pattern books at secondhand book shops.

Marking a stitch
- Here's a quick and easy marker: Cut a one-eighth-inch slice off a plastic drinking straw. The straw rings fit right on knitting needles up to size ten.
- Or, try using a twist tie. It's easy to transfer from needle and works just as well as expensive plastic markers.

Stitch-holder substitutes
- Single bar-type barrettes are great for this.
- Or, use a metal shower curtain ring.
- Or use safety pins but put a small button on the pin first so yarn doesn't get caught in the spring end.

Substitutes for cable needles
- Use a bobby pin.
- Or hair pin.

Substitute counter
- Use a cribbage board to count the rows.

Pick up stitches
- If you don't have a crochet hook to pick up dropped stitches, try a nutpicker.

Keeping stitches on the needles
- When you lay work aside, wind a rubber band tightly around the tips of the needles and tuck the first stitch under it for safekeeping.
- Or press small corks on the tips.

Sewing a sweater together
- Instead of attaching the pieces with straight pins, use plastic picks that come with hair rollers. They're much longer, stronger and won't get lost in the sweater.

Fringe benefits
● Double the number of fringes you make at one time by winding yarn around a magazine lengthwise instead of a smaller object. Cut yarn top and bottom, then double over each bunch to make a fringe.

Darn it
● Slip crocheted bedroom slippers over a shoe to make darning worn spots lots easier.

Storing large quantities of yarn
● If you've bought an awful lot of yarn to make bedspreads or afghans, store in cloth laundry bags or trash bags.
● Or, use liquor cartons with compartment dividers. Colors or types of yarn can be kept separately.

Let's Get Packing

Advance planning
- Pack last what you'll be using first: dinner dress, nightgown, accessories, bathroom stuff.
- Put a change of clothing in your husband's bag while he packs a change in yours. Then if one bag is lost or delayed in transit, you'll both have something clean to wear.

Do-it-yourself travel kits
- Fill a plastic egg-shaped panty hose container with your jewelry. It's compact and closes tightly so contents won't spill out into suitcase.
- Turn an empty 35-mm film cannister into a nifty sewing kit. Insert a few buttons, bobbins of thread, needles, safety pins and so on.
- Protect cosmetics and small bottles by putting them in a small plastic lunchbox to tuck in your luggage.

Avoiding spills
- Squeeze a well-filled bottle until the contents are almost at the rim, then twist the cap on quickly. The powerful suction will keep your moisturizer or shampoo from squirting or dripping all over your cosmetic case.

Case closed
- Identification tags that hang on the outside of the case can come off easily. For additional protection, place a business card or your name and address inside the liner of the suitcase. Then, also tape on the outside. Cover completely with tape so it can't be pulled off.
- Tote bags will travel better if you tie both handles together with twine. The bag will be less likely to rip open during handling.

Spot your luggage quickly
- Just wrap brightly colored electrical tape around the handle.
- Or tie on some colorful yarn.

Traveling abroad
- Make note of your passport and traveler's check numbers and carry them in a separate place from your passport and checks.
- Carry along a bottle opener. It'll come in handy to open bottles you purchase in a train station or grocery store.
- Take single United States dollar bills for tipping in case you don't have local currency at the beginning or end of your trip.
- Find out the local exchange rate and before you go shopping, make yourself a chart of the equivalent of one dollar, five dollars, ten dollars, etc., for quick reference.

On the Road

Settling the seating dispute
- When kids fight about where to sit, have them draw straws to see who gets what seat.

Portable art cases
- A covered 9 × 13 inch cake pan is great for holding crayons, pencils, paper and small toys for small children. Closed, the pan makes a good drawing board.
- Or, hang a shoe bag on the coat hook in the car and fill the pouches with small toys, crayons, drawing pads, etc.

For a change of pace
- Play their favorite records or stories on cassettes on a portable tape recorder.

Baby business
- To keep a bottle chilled, put it in a wide-mouthed insulated Thermos bottle with a few chunks of ice.
- Or, put it inside a locking plastic bag with a few ice cubes. Wrap it in a diaper or dish towel.

Taking your child's friend along
- Have the parents give you a certified letter (from a notary public) agreeing to emergency medical treatment for the child in case of accident or illness.

Moving to a New House

Get organized

- Try a color coding system. Use a different colored marking pen for each room. In the new house, pin up pieces of colored paper to the door of each room so the movers know where to drop the corresponding cartons.
- Buy a small spiral notebook to keep track of boxes. Number each box, then note each number down, also recording which box goes where. You'll have a handy reference book when deciding which box to unpack first.

Thinking ahead

- If you're moving to another state or city at some distance, be smart — pack enough clothing for the first five days for everyone in the family. If the movers are late, you won't be caught without a change.

Packing

- Lawn and leaf bags are great to pack blankets, clothes and other soft items. They're less expensive than boxes and can be used to store items that are not needed right away.
- Extra-large plastic garbage cans are also great packing containers. Stuff them with unbreakables or pack them with breakables that have been wrapped carefully in padding.
- Use small linens such as towels, washcloths and pillowcases for packing material for dishes. They protect dinnerware and don't take up extra space.
- Press screws and other hardware to a length of duct tape, then press the tape to the back of the item for which they're needed. Hardware won't get lost in transit. (But be careful where you stick it. Duct tape can remove some finishes.)

Total recall

- Take a photograph of the arrangement of a complicated display of china and glassware in your cabinets. When you unpack in the new house, you'll have a handy guide for reassembly.

The Great Outdoors

Ice chests
- Prechill the interior by filling the chest with ice. Let it stand for at least fifteen minutes, then drain and cover the bottom with plenty of fresh ice cubes.
- Or, if you've got the space, put the chest directly in the freezer.

Transporting wine bottles and glasses
- Wrap breakables in double thicknesses of paper towel.

To hang a roll of paper towels outside
- It's a cinch when you use a wire hanger that's got a narrow paper tube bottom bar. Just pull the wire out of one end of the paper tube, insert the roll of toweling, then push the wire back in the tube.

Just whistle an SOS
- Carry along a whistle when you head for the woods. If you should become lost, blow it sharply three times and repeat this pattern every few minutes. Any kind of signal repeated three times at frequent intervals is a universal call and should bring help.

Mini ovens
- Roast potatoes on an open fire, but drop them into an empty tin can so they won't char.

Improvised barbecue grill
- Line the inside of a metal garbage can lid with aluminum foil, then prop on four empty beer or soda cans. If you've got bricks, even better. Use instead of the cans. Drop in some charcoal and place a grate on top.

Cleaning cast iron
- To clean without using water, wipe out the pan first with a paper towel. Pour a tablespoon or two of salt into the skillet, then scrub the pan clean with another paper towel. Salt will absorb grease while acting like scouring powder. Wipe out remaining salt.

Grit-free soap
- Slip a bar of soap into a nylon stocking to prevent any sand from getting on the soap when bathing in the lake.

Substitute clotheslines
- Fold a long rope in half and wrap the middle of it around a tree. Hold the two ends together and twist them around a tree. Tie the ends around another tree. To hang clothes, just slip a piece of them between the double twisted rope. You don't need pins.
- Or, try an old pair of panty hose. The nylon stretches to about 2.7 yards.

Dry storage in your boat
- Protect matches and other items from dampness in an open boat by stashing them in screwtop plastic jars mounted under the seats.

At the beach with baby
- A big sheet will be cooler than a blanket for you and baby to sit on. It shakes clean easier, too.
- Fill a mesh bag with beach toys. It can all be dunked in the water to clean and drain dry.

By the way
- Share your favorite original hints with us. If we use them we will send you a copy of Mary Ellen's Kitchen Hints. Send to: Mary Ellen, 6414 Cambridge Street, St. Louis Park, Minnesota 55426

Index

153

Also available from Doubleday
Mary Ellen Family Books™

BABY'S FIRST HELPINGS
Superhealthy Meals for Superhealthy Kids
Chris Casson Madden
ISBN: 0-385-19143-X   comb-bound paperback $5.95

TAKING CARE OF MOMMY
Everything Your Doctor Wouldn't, Your Mother Couldn't, and Your Best Friend Didn't Tell You About Becoming a Mommy
Susan G. Gross and Paula S. Linden, R.D.
ISBN: 0-385-19144-8   paperback $7.95

THE MILLION DOLLAR CONTEST COOKBOOK
Jean Sanderson
ISBN: 0-385-19145-6   comb-bound paperback $6.95

GREAT BEGINNINGS, GREAT ENDINGS
Make Ahead Hors d'oeuvres and Desserts
Carol Sakstrup and Jane Charboneau
ISBN: 0-385-19146-4   comb-bound paperback $3.95

GIFT OF FRIENDSHIP COUPONS
Patti Brietman and Nanc Gordon
ISBN: 0-385-19218-5   paperback $3.95

500 HINTS FOR KIDS
Randy Harleson and Eileen Cavanaugh
ISBN: 0-385-19217-7   paperback $5.95

All of the Mary Ellen Family Books™ are available now at your local bookseller.